I0881781

Topics in Number Theory

Gail Gallitano and Shiv Gupta

Department of Mathematics

West Chester University

Kendall Hunt
publishing company

Cover image © Shutterstock, Inc.

Kendall Hunt
publishing company

www.kendallhunt.com
Send all inquiries to:
4050 Westmark Drive
Dubuque, IA 52004-1840

Copyright © 2018 by Kendall Hunt Publishing Company

ISBN 978-1-5249-4801-6

All rights reserved. No part of this publication may be reproduced, stored in a retrieval system, or transmitted, in any form or by any means, electronic, mechanical, photocopying, recording, or otherwise, without the prior written permission of the copyright owner.

Published in the United States of America

Dedicated to the Memory of

Our Dear Friend and Colleague

Professor Waclaw Szymanski (1949–2016)

PhD, DSc (Polish Academy of Sciences)

Contents

Introduction and Acknowledgment

Number theory is an ancient and exciting branch of pure mathematics which is devoted to the study of integers. Carl Gauss (1777–1855), sometimes referred to as the "greatest mathematician since antiquity," once stated that "Mathematics is the queen of the sciences and number theory is the queen of mathematics." Many mathematicians, both those BCE and those of the CE, such as Plato, Euclid, Diophantus, Archimedes, Eratosthenes Aryabhata, Brahmagupta, Bhaskara, Fermat, Euler, Gauss, Lagrange, Legendre, Ramanujan, and others, studied and contributed to the field of number theory with some devoting their entire life to the study of this magnificent and vibrant subject.

It is conjectured that Kronecker remarked, "God created the natural numbers and all the rest is the work of man." It is our experience that undergraduate students equipped only with the mathematics they learned in high school respond rather well to the teaching of number theory. It is for these reasons that we have consistently advocated for a required course in number theory for undergraduate mathematics majors at West Chester University and elsewhere. We hope that these notes on number theory will strengthen this case and convince our colleagues and others of the desirability of making a course on number theory a required course for undergraduate mathematics majors at our university and elsewhere.

The first author had the opportunity to teach several topics in number theory as part of several courses over the 45 plus years of her teaching career. She taught at various institutions including Fairleigh Dickinson University, Teaneck, NJ, Montclair University, Montclair, NJ, Shippensburg University, Shippensburg, PA, The University of Nevada, Las Vegas, NV, and West Chester University, West Chester, PA. The second author had the opportunity to teach the theory of numbers at various institutions, including St. Stephens College, Delhi, India, the University of Jos in Nigeria, and West Chester University, West Chester, PA. At West Chester University he also taught a seminar course on Computational Number Theory. Among other topics, the contents of chapter 10 of these notes were discussed in this course.

In the bibliography, we have given a list of several books on number theory that we have consulted, at one time or another, either while teaching or while preparing these notes. Apart from the choice of material and some personal preferences for presentation of certain proofs, no claim to any originality is asserted. There may be some omissions, misprint, and even mistakes in this book. We would of course appreciate learning about them from the readers of these notes and will acknowledge them promptly. It is hoped that readers will find something of interest in our book. Finally, we would like to thank all of the students at various institutions who gave us an opportunity to teach them something about number theory, a subject that we hold very dear to our hearts.

—Gail Gallitano and Shiv Gupta

West Chester, Pennsylvania

Prerequisites and Notation

In this book, we assume that the reader is familiar with the basic concepts of set theory, equivalence relations, and equivalence classes. We also assume familiarity with elementary abstract algebra—groups, subgroups, and their basic properties (Abelian groups, order of an element of a group, Lagrange's theorem, etc.), rings, subrings, ideals, integral domains fields, vector spaces, and so on. We use the following notations:

$\mathbb{N}$ = Set of all natural numbers = $\{1, 2, \cdots\cdots\}$.

$\mathbb{Z}$ = Set of all integers.

$\mathbb{Q}$ = Set of all rational numbers.

$\mathbb{R}$ = Set of all real numbers.

$\mathbb{C}$ = Set of all complex numbers.

C_n = Cyclic group of order n.

$\Phi_n(x)$ = nth cyclotomic polynomial.

$a|b$ = a divides b.

$a \nmid b$ = a does not divides b.

GCD, or gcd = Greatest Common Divisor.

LCM or lcm = Least common multiple.

LHS = Left-hand side.

RHS = Right-hand side.

$\mathbb{Z}_n$ or $\mathbb{Z}/n\mathbb{Z}$ = Ring consisting of equivalence classes modulo n.

$\mathbb{Z}_n^*$ = Multiplicative group of the invertible elements in $\mathbb{Z}_n$.

$\emptyset$ = Empty set.

$e^{ix} = \cos x + i \sin x$.

We often use the term "number" rather than the precise term "integer" or whole number. In any case, it will be clear from the context what is meant and no confusion will result.

CHAPTER 1

Generalities

1. Divisibility

Let a and b be two integers, $a \neq 0$. We shall say that a divides b (written as $a|b$) if $b = ac$, for some integer c. If a divides b then we shall also say that b is a multiple of a. The following properties of division can be easily verified.

(i) $a|0$ for all $a \neq 0$ in $\mathbb{Z}$.

(ii) $a|b$, $b|c$ implies that $a|c$ for all a, b and c in $\mathbb{Z}$.

(iii) $a|b$, $a|c$ implies that $a|(bx + cy)$, for all a, b, c, x, y in $\mathbb{Z}$.

Greatest Common Divisor

Let a and b, not both zero be two integers. A positive integer d is said to be Greatest Common Divisor (GCD) of a and b if

(i) $d|a$, $d|b$.

(ii) If $0 \neq d_1 \in \mathbb{Z}$ is such that $d_1|a$ and $d_1|b$ then $d_1|d$.

The GCD of two integers (not both zero) exists and is the largest positive integer which divides each of the two integers and is therefore uniquely determined.

If d is the GCD of a and b we shall write $GCD(a, b) = d$. If $GCD(a, b) = 1$ then we shall say that a and b are relatively prime or coprime.

Least Common Multiple

Let a and b be two nonzero integers. A positive integer m is said to be LCM (Least Common Multiple) of a and b if:

(i) $a|m$, $b|m$.

(ii) If $0 \neq m_1 \in \mathbb{Z}$ is such that $a|m_1$ and $b|m_1$ then $m|m_1$.

The LCM of two nonzero integers exists and is the smallest positive integer which is divisible by each of the two integers. If m is the LCM of a and b then we shall write $LCM(a, b) = m$.

Notes

(1) We have only defined the notion of GCD and LCM of two integers. These notions can be easily extended to the GCD and LCM of more than two integers.

(2) If a and b are two positive integers then it can be easily seen that

$$GCD(a,b) \cdot LCM(a,b) = a \cdot b$$

2. Division Algorithm

Let a and b be two integers, $b > 0$. Then there exist unique integers q and r such that $a = b \cdot q + r$, where $0 \leq r < b$. So the *Division Algorithm* is simply a restatement of the apparently obvious fact that when an integer is divided by a positive integer then there is a quotient and a remainder. The Euclidean algorithm is a method of finding the GCD of two integers, not both zero, by the repeated use of the Division Algorithm. The implementation of the Euclidean algorithm on a computer provides a good example of the so-called recursive programming technique. It is based on the fact that (in the above notation) $GCD(a,b) = GCD(bq+r,b) = GCD(r,b) = GCD(b,r)$.

Let a and b (not both zero) be two integers. If $GCD(a,b) = d$ then with the help of Euclidean Algorithm we can find integers x and y (not unique) such that $ax + by = d$.

The existence of such an x and y can, however, be deduced easily from the fact that a nonempty set of positive integers has a least element. Let

$$\Omega = \{a\alpha + b\beta : \alpha, \beta \in \mathbb{Z}, a\alpha + b\beta \geq 0\}$$

Then $\Omega \neq \emptyset$. Let d be the least member of Ω. So $d = ax + by$ for some $x, y \in \mathbb{Z}$. Then clearly $GCD(a,b) = d$. In fact, it can be shown that every element of Ω is a multiple of d. In particular it follows that if a and b are relatively prime (that is if $GCD(a,b) = 1$) then there exist integers x and y such that $ax + by = 1$. We shall often use this in the text.

The following example will illustrate the computation of GCD of two numbers by successive application of the Division Algorithm.

An Example

Let $a = 15625$, and $b = 1024$. We have

$$\begin{aligned} 15625 &= \mathbf{15} \cdot 1024 + 265 \\ 1024 &= \mathbf{3} \cdot 265 \quad + 229 \end{aligned}$$

$$
\begin{aligned}
265 &= \mathbf{1} \cdot 229 + 36 \\
229 &= \mathbf{6} \cdot 36 \ \ + 13 \\
36 &= \mathbf{2} \cdot 13 \ \ + 10 \\
13 &= \mathbf{1} \cdot 10 \ \ + \ \ 3 \\
10 &= \mathbf{3} \cdot 3 \ \ \ + \ \ 1 \\
3 &= \mathbf{1} \cdot 3 \ \ \ + \ \ 0
\end{aligned}
$$

The last nonzero remainder namely 1 is the GCD of 15625 and 1024. The sequence of operations carried out above ending in computing the GCD of two numbers is called the *Euclidean Algorithm.* We can express 1 as a linear combination of 15625 and 1024 as follows:

$$
\begin{aligned}
1 &= 10 - 3 \cdot 3 \\
&= 10 - (13 - 1 \cdot 10) \cdot 3 \\
&= 4 \cdot 10 - 3 \cdot 13 \\
&= 4(36 - 2 \cdot 13) - 3 \cdot 13 \\
&= 4 \cdot 36 - 11 \cdot 13 \\
&= 4 \cdot 36 - 11 \cdot (229 - 6 \cdot 36) \\
&= 70 \cdot 36 - 11 \cdot 229 \\
&= 70(265 - 1 \cdot 229) - 11 \cdot 229 \\
&= 70 \cdot 265 - 81 \cdot 229 \\
&= 70 \cdot 265 - 81(2024 - 3 \cdot 265) \\
&= 313 \cdot 265 - 81 \cdot 1024 \\
&= 313(15625 - 15 \cdot 1024) - 81 \cdot 1024 \\
&= 313 \cdot 15625 - 4776 \cdot 1024
\end{aligned}
$$

So $1 = 15625x + 1024y$, where $x = 313$ and $y = -4776$. It is easy to see that x and y are not unique. These x and y can also be found somewhat differently. As we noticed the last nonzero remainder, namely 1 in the successive application of the division algorithm above appeared in the seventh step. We define seven 2×2 matrices A_i, $1 \leq i \leq 7$ as follows. Let q_i, $1 \leq i \leq 7$, denote the successive quotients in the seven steps in the application of division algorithm above. That is

$$q_1 = 15, \quad q_2 = 3, \quad q_3 = 1, \quad q_4 = 6, \quad q_5 = 2, \quad q_6 = 1, \quad q_7 = 3.$$

Then

$$A_i = \begin{bmatrix} 0 & 1 \\ 1 & -q_i \end{bmatrix}, \qquad 1 \leq i \leq 7.$$

Therefore

$$A_1 = \begin{bmatrix} 0 & 1 \\ 1 & -15 \end{bmatrix}, \quad A_2 = \begin{bmatrix} 0 & 1 \\ 1 & -3 \end{bmatrix}, \quad A_3 = \begin{bmatrix} 0 & 1 \\ 1 & -1 \end{bmatrix}, \quad A_4 = \begin{bmatrix} 0 & 1 \\ 1 & -6 \end{bmatrix},$$

$$A_5 = \begin{bmatrix} 0 & 1 \\ 1 & -2 \end{bmatrix} A_6 = \begin{bmatrix} 0 & 1 \\ 1 & -1 \end{bmatrix}, \quad A_7 = \begin{bmatrix} 0 & 1 \\ 1 & -3 \end{bmatrix}.$$

The first seven steps of the division algorithm in the computation of GCD of 15625 and 1024 can be represented (respectively) as the following seven matrix equations:

$$\begin{bmatrix} 0 & 1 \\ 1 & -15 \end{bmatrix} \begin{bmatrix} 15625 \\ 1024 \end{bmatrix} = \begin{bmatrix} 1024 \\ 265 \end{bmatrix}, \quad \begin{bmatrix} 0 & 1 \\ 1 & -3 \end{bmatrix} \begin{bmatrix} 1024 \\ 265 \end{bmatrix} = \begin{bmatrix} 265 \\ 229 \end{bmatrix},$$

$$\begin{bmatrix} 0 & 1 \\ 1 & -1 \end{bmatrix} \begin{bmatrix} 265 \\ 229 \end{bmatrix} = \begin{bmatrix} 229 \\ 36 \end{bmatrix}, \quad \begin{bmatrix} 0 & 1 \\ 1 & -6 \end{bmatrix} \begin{bmatrix} 229 \\ 36 \end{bmatrix} = \begin{bmatrix} 36 \\ 13 \end{bmatrix}$$

$$\begin{bmatrix} 0 & 1 \\ 1 & -2 \end{bmatrix} \begin{bmatrix} 36 \\ 13 \end{bmatrix} = \begin{bmatrix} 13 \\ 10 \end{bmatrix}, \quad \begin{bmatrix} 0 & 1 \\ 1 & -1 \end{bmatrix} \begin{bmatrix} 13 \\ 10 \end{bmatrix} = \begin{bmatrix} 10 \\ 3 \end{bmatrix},$$

$$\begin{bmatrix} 0 & 1 \\ 1 & -3 \end{bmatrix} \begin{bmatrix} 10 \\ 3 \end{bmatrix} = \begin{bmatrix} 3 \\ 1 \end{bmatrix}.$$

As

$$A_7 \times A_6 \times A_5 \times A_4 \times A_3 \times A_2 \times A_1 = \begin{bmatrix} -85 & 1297 \\ \mathbf{313} & \mathbf{-4776} \end{bmatrix},$$

we have

$$\begin{bmatrix} -85 & 1297 \\ \mathbf{313} & \mathbf{-4776} \end{bmatrix} \begin{bmatrix} 15625 \\ 1024 \end{bmatrix} = \begin{bmatrix} 3 \\ 1 \end{bmatrix}.$$

Which in particular gives

$$\mathbf{313} \cdot 15625 + (\mathbf{-4776}) \cdot 1024 = 1.$$

3. Primes and Composites

A positive integer $p > 1$ is said to be a *prime number* or simply a *prime* if it has no (positive) divisors other than 1 and p itself. One (1) itself, by definition, is not a prime. The first eight primes are $2, 3, 5, 7, 11, 13, 17, 19$. A positive integer $n > 1$ is said to be *composite* if it is not a prime number. A consequence of the definition of a prime number is that if a prime number p divides the product $a \cdot b$ then either p divides a or p divides b. For, if p does not divide a then $GCD(a, p) = 1$ and we can find $x, y \in \mathbb{Z}$ such that $ax + py = 1$. Multiplying this equation by b we get $axb + pyb = b$. Since p divides ab, there exists $q \in \mathbb{Z}$ such that $ab = pq$. Substituting pq for ab in the equation $axb + pyb = b$, we get $pqx + pyb = b$ or $p(qx + yb) = b$ which implies that p divides b. This property of a prime number, namely if a prime p divides the product ab then it divides a or b implies the so-called **Fundamental Theorem of Arithmetic**, namely the fact that every natural number can be uniquely expressed (apart from order) as a product of prime numbers.

Notes

(1) Mathematicians since the time of antiquity have been fascinated with prime numbers. As every natural number can be factored into primes, these constitute the "building blocks" of all numbers. In as much as all compounds in nature can be thought of being made up from various elements, all numbers are indeed "made up" from prime numbers. Most students who learn group theory are (unfortunately) unaware of the fact that the unique factorization property of integers (*viz.* Fundamental Theorem of Arithmetic) is the motivation behind the Jordan Holder Theorem in group theory concerning the composition factors of a finite group.

(2) Euclid proved that the sequence of primes does not end. In fact, the idea behind Euclid's proof is very simple. Euclid observed that if $p_1, p_2, p_3, \cdots, p_n$ are primes then the number $N = p_1p_2p_3 \cdot p_n + 1$ is *not* divisible by any of these primes. So N is divisible by a prime number different from $p_1, p_2, p_3, \cdots, p_n$.

(3) A composite number n must have a prime factor $\leq \sqrt{n}$. This is an immediate consequence of the unique factorization property. This observation provides a method of testing primality of not-too-large numbers. In order to test whether a given number is a prime or not we simply have to check whether n is divisible by any of the primes $\leq \sqrt{n}$. If n is not divisible by any of the primes $\leq \sqrt{n}$ then n is a prime number.

(4) Sieve of Eratosthenes provides a method for making a list of prime numbers below a given number n. This works as follows:

(a) Write the numbers from 2 to n.

(b) Circle the number 2 and cross out (or blacken out) every second number, namely, $4, 6, 8, \cdots$ all the way up to n. In another words cross out the multiples of 2 (other than 2 itself) up to n.

(c) Circle the first uncrossed number and cross out every number occurring at the interval determined by the most recently circled number. For example, if we have circled the number k then we cross out the numbers $2k, 3k, \cdots$, etc.

(d) Continue the above process for the numbers up to the largest integer not exceeding $\sqrt{n}$. After this circle all the numbers that are not crossed out.

(e) The numbers which have been circled are all the prime numbers below n.

We shall illustrate the method of Sieve of Eratosthenes by finding primes below 50.

- First we write the natural numbers from 2 to 50. As $7 < \sqrt{50} < 8$, we boldface the number 7 indicating that our sieving process will stop after we have sieved-out all the multiples of 7.

	2	3	4	5	6	**7**	8	9	10
11	12	13	14	15	16	17	18	19	20
21	22	23	24	25	26	27	28	29	30
31	32	33	34	35	36	37	38	39	40
41	42	43	44	45	46	47	48	49	50

- We circle 2 and blacken-out all the remaining multiples of 2.

	②	3	■	5	■	**7**	■	9	■
11	■	13	■	15	■	17	■	19	■
21	■	23	■	25	■	27	■	29	■
31	■	33	■	35	■	37	■	39	■
41	■	43	■	45	■	47	■	49	■

- We circle the number 3 and blacken out all the remaining multiples of 3, namely 9, 15, 21, 27, 33, 39, and 45.

	②	③	■	5	■	7	■	■	■
11	■	13	■	■	■	17	■	19	■
■	■	23	■	25	■	■	■	29	■
31	■	■	■	35	■	37	■	■	■
41	■	43	■	■	■	47	■	49	■

- We circle the number 5 and blacken out all the remaining multiples of 5, namely 25 and 35.

	②	③	■	⑤	■	7	■	■	■
11	■	13	■	■	■	17	■	19	■
■	■	23	■	■	■	■	■	29	■
31	■	■	■	■	■	37	■	■	■
41	■	43	■	■	■	47	■	49	■

- Finally we circle the number 7 and blacken-out the only remaining multiple of 7, namely 49.

	②	③	■	⑤	■	⑦	■	■	■
11	■	13	■	■	■	17	■	19	■
■	■	23	■	■	■	■	■	29	■
31	■	■	■	■	■	37	■	■	■
41	■	43	■	■	■	47	■	■	■

- We circle the remaining numbers (which have not been blackened out). These circled numbers are all the primes below 50.

	②	③	■	⑤	■	⑦	■	■	■
⑪	■	⑬	■	■	■	⑰	■	⑲	■
■	■	㉓	■	■	■	■	■	㉙	■
㉛	■	■	■	■	■	㊲	■	■	■
㊶	■	㊸	■	■	■	㊼	■	■	■

So the primes below 50 are $2, 3, 5, 7, 11, 13, 17, 19, 23, 29, 31, 37, 41, 43$, and 47.

Various modifications have been suggested in the method of the Sieve of Eratosthenes with a view to improve its efficiency. We would like to mention a paper by Xuedong Luo, "A practical sieve algorithm for finding prime numbers," *Communications of the ACM,* March 1989, Vol 32, number 3, pages 344-346, where the author discusses an extension of the method of the Sieve of Eratosthenes which is more suitable for finding primes below a large natural number.

(5) Because of their use in cryptography, especially RSA (Rivest, Shamir, and Adleman) cryptography (which we shall discuss in chapter 10) prime numbers and primality testing have been of much interest to computer scientists and mathematicians interested in computational number theory. We shall also discuss some methods of testing primality of a large natural numbers in the same chapter.

4. Congruence and Their Properties

Let a, b, m be integers, $m > 0$. We shall say that a is congruent to b mod m (written as $a \equiv b \pmod{m}$ or simply $a \equiv b(m)$) if $a - b$ is a multiple of m, that is, $m|(a - b)$.
The following properties of congruence can be easily deduced from the divisibility properties of integers.

(i) $a \equiv a \pmod{m}$.

(ii) If $a \equiv b \pmod{m}$, then $b \equiv a \pmod{m}$.

(iii) If $a \equiv b \pmod{m}$ and $b \equiv c \pmod{m}$, then $a \equiv c \pmod{m}$.

(iv) Let $c, d \in \mathbb{Z}$. If $a \equiv c \pmod{m}$ and $b \equiv d \pmod{m}$, then $a \pm b \equiv c \pm d \pmod{m}$ and $a \cdot b \equiv c \cdot d \pmod{m}$. In particular, $a \equiv b \pmod{m} \Longrightarrow a^k \equiv b^k \pmod{m}$ for $k \in \mathbb{Z}$, $k \geq 0$.

The properties (i), (ii), and (iii) above imply that for given positive integer m the relation $a \sim b$ if $a \equiv b \pmod{m}$ is an equivalence relation on $\mathbb{Z}$. The set of equivalence classes under this equivalence relation is denoted by $\mathbb{Z}/(m)$ or $\mathbb{Z}/m\mathbb{Z}$, or simply $\mathbb{Z}_m$. We shall often use the notation $\mathbb{Z}_m$. The equivalence class mod m (i.e., an element of $\mathbb{Z}_m$) corresponding to the element $a \in \mathbb{Z}$ will be denoted by $\bar{a}$. In $\mathbb{Z}_m$ we define "addition" and "multiplication" as follows:

$$\bar{a} + \bar{b} = \overline{a + b}, \qquad \bar{a} \cdot \bar{b} = \overline{a \cdot b}$$

In view of the properties of congruence it is easily seen that the addition and multiplication in $\mathbb{Z}_m$ are well defined and these make $\mathbb{Z}_m$ into a commutative ring with identity (namely $\bar{1}$). If n is any integer and m a positive integer then by the division algorithm we can write $n = q \cdot m + r$, where $0 \leq r < m$. So $n \equiv r (\text{mod } m)$. Thus any number n is congruent modulo m to one of the numbers in the set $\{0, 1, 2 \cdots, m-1\}$. Also no two distinct members of the set $\{0, 1, 2 \cdots, \ m-1\}$ are congruent mod m. As noted earlier, if $a, b \in \mathbb{N}$ with $GCD(a, b) = 1$ then there exist $x, y \in \mathbb{Z}$ such that $ax + by = 1$. This implies that $ax \equiv 1$ (mod b) and $by \equiv 1$ (mod a).

Beginning students learning about the concept of congruence have some difficulty with it. The following illustration may help them to understand this concept somewhat better. Let $m = 8$ and we shall discuss the meaning of $a \equiv b$ (mod 8). We arrange *all* the integers in eight columns as follows:

⋮	⋮	⋮	⋮	⋮	⋮	⋮	⋮
−48	−47	−46	−45	−44	−43	−42	−41
−40	−39	−38	−37	−36	−35	−34	−33
−32	−31	−30	−29	−28	−27	−26	−25
−24	−23	−22	−21	−20	−19	−18	−17
−16	−15	−14	−13	−12	−11	−10	−9
−8	−7	−6	−5	−4	−3	−2	−1
0	**1**	**2**	**3**	**4**	**5**	**6**	**7**
8	9	10	11	12	13	14	15
16	17	18	19	20	21	22	23
24	25	26	27	28	29	30	31
32	33	34	35	36	37	38	39
40	41	42	43	44	45	46	47
48	49	50	51	52	53	54	55
⋮	⋮	⋮	⋮	⋮	⋮	⋮	⋮

We note that any two numbers in the same column differ by a multiple of eight, that is, their difference is a multiple of eight. Let $a, b \in \mathbb{Z}$. Then following three statements have essentially the same meaning

- a and b lie in the same column in the above table.

- a and b leave same remainder when divided by 8.
- $a \equiv b \pmod{8}$.

We can make such columns for any positive integer m but it would be inconvenient and unnecessary to do so for large m. If we pick up one member from each column in the above table then the resulting eight numbers would constitute what is known as a complete residue system modulo 8. The set of all the entries in a column constitute a congruence class modulo 8 and so there are eight congruence classes modulo 8 and each of these congruence classes can be represented by any of its member. Suppose that we are asked to find all $x \in \mathbb{Z}$ satisfying $5x \equiv 7 \pmod{8}$. Such "equations" are called congruence in number theory. If a is a solution of this congruence then any number lying in the same column as a in the above table is also a solution. In fact, if both a and b are solutions of the congruence $5x \equiv 7 \pmod{8}$ then $a \equiv b \pmod{8}$. Thus, as $a = 3$ is a solution, therefore, every integer of the form $8k + 3$ is also a solution of the congruence $5x \equiv 7 \pmod{8}$.

5. Complete Residue System mod m

Let m be a positive integer. Then a set of m numbers $\{x_1, x_2, \cdots, x_m\}$ is called *complete residue system* mod m if no two distinct members of the set are congruent modulo m. It is clear that if $x_1, x_2, \cdots, x_m$ is a complete residue system mod m, and we replace each member of this set by its least nonnegative residue mod m (that is the least nonnegative number in the corresponding column for m) then we shall get the set $\{0, 1, \cdots, m-1\}$. In fact, any set of m consecutive integers constitutes a complete residue system modulo m.

6. Chinese Remainder Theorem

The following theorem which was known to the Chinese over 2000 years ago is very useful in solving congruence.

Theorem Let $m_1, m_2, \cdots, m_k$ be positive integers which are pair-wise coprime (i.e., $gcd(m_i, m_j) = 1$, if $i \neq j$) and $a_1, a_2, \cdots, a_k$ be any integers. Then there exists an integer x such that $x \equiv a_i (\text{mod } m_i)$, $1 \leq i \leq k$. Moreover, such an x is uniquely determined modulo $M = m_1 m_2 \cdots m_k$.

Proof: Let $M = m_1 \cdot m_2 \cdots m_k$, and $M_i = \dfrac{M}{m_i}$, $1 \le i \le k$. Evidently $gcd(m_i, M_i) = 1$. Choose b_i, $1 \le i \le k$ such that $b_i M_i \equiv 1 \pmod{m_i}$. Let

$$x = \sum_{i=1}^{k} a_i b_i M_i$$

Then $x \equiv a_i (\text{mod } m_i)$, for $1 \le i \le k$. Suppose that y is such that $y \equiv a_i (\text{mod } m_i)$, $1 \le i \le k$. Then $x \equiv y \pmod{m_i}$, that is $x - y \equiv 0 (\text{mod } m_i)$, $1 \le i \le k$. Since $gcd(m_i, m_j) = 1$ for $i \ne j$, it follows that $x \equiv y \pmod{m_1 \cdot m_2 \cdots m_k}$, that is $x \equiv y \pmod{M}$. This completes the proof of the Chinese Remainder Theorem.

An Example

Let us find the smallest natural number x satisfying the following conditions:

- when x is divided by 3 remainder is 2.
- when x is divided by 5 remainder is 3.
- when x is divided by 7 remainder is 4.
- when x is divided by 8 remainder is 5.

This means that we have to solve the following four simultaneous congruence:

$$x \equiv 2 \pmod{3}, \quad x \equiv 3 \pmod{5}, \quad x \equiv 4 \pmod{7}, \quad x \equiv 5 \pmod{8}$$

To solve this problem we let $k = 4$, $m_1 = 3$, $m_2 = 5$, $m_3 = 7$, $m_4 = 11$, $a_1 = 2$, $a_2 = 3$, $a_3 = 4$, $a_4 = 5$. Also $M = 840$ and $M_1 = 280$, $M_2 = 168$, $M_3 = 120$, $M_4 = 105$ and $b_1 = 1$, $b_2 = 2$, $b_3 = 1$ and $b_4 = 1$. The last column of the following table gives the value of $a_i b_i M_i$ for $1 \le i \le 4$.

i	m_i	M_i	$M_i(\text{mod } m_i)$	b_i	a_i	$a_i b_i M_i$
1	3	280	1	1	2	560
2	5	168	3	2	3	1008
3	7	120	1	1	4	480
4	8	105	1	1	5	425

Therefore, $x = \sum_1^4 a_i b_i M_i = 560 + 1008 + 480 + 525 = 2573 \equiv 53 \pmod{840}$.

CHAPTER 2

Some Number Theoretic Functions

Arithmetical Function

A real or complex valued function whose domain is the set of natural numbers is called an **arithmetical function** or a **number-theoretical function**. An arithmetical function f is called *multiplicative* if $f(mn) = f(m)f(n)$ whenever $gcd(m, n) = 1$ It is called *totally multiplicative* or *completely multiplicative* if $f(mn) = f(m)f(n)$ holds for all natural numbers m and n. In this chapter we shall study the following number-theoretical functions:

(1) The Euler's ϕ-function (or Euler's Totient function).

(2) The Greatest Integer Function.

(3) The function $\tau(n)$ - the number of divisors of n.

(4) The function $\sigma(n)$ - the sum of the divisors of n.

(5) Mobius μ-function.

1. Euler's ϕ-function

Let m be a positive integer. The number of positive integers less than or equal to m and co-prime to m is denoted by $\phi(m)$. This function ϕ is known as Euler's ϕ-function or Euler's totient function. So $\phi(m)$ is simply the cardinality of the set $\{i : 1 \leq i \leq m, gcd(i, m) = 1\}$.

Remark

Since $x \equiv y(\text{mod } m)$ implies that $gcd(x, m) = gcd(y, m)$ it follows that if $\{x_1, x_2, \cdots, x_m\}$ is any complete residue system modulo m then $\phi(m)$ can be defined as the cardinality of the set $\{x_i : 1 \leq i \leq m, gcd(x_i, m) = 1\}$. It is clear that p is a prime number if and only if $\phi(p) = p - 1$. Also, for a prime power p^k, $\phi(p^k) = p^k - p^{k-1}$, as the only numbers less than or equal to p^k that are *not* co-prime to p^k are multiples of p which are less than or equal to p^k. Since there are precisely p^{k-1} such multiples of p, it follows that $\phi(p^k) = p^k - p^{k-1}$.

The following important theorem helps us to compute the value of $\phi(m)$ once the factorization of m as a product or primes is known.

Theorem: If m and n are positive integers such that $gcd(m,n) = 1$ then $\phi(m \cdot n) = \phi(m) \cdot \phi(n)$.

Proof :

We first arrange the numbers from 1 to mn in m rows and n columns as follows:

$$\begin{array}{cccccc} 1, & 2, & 3, & \cdots, & \cdots, & n \\ n+1, & n+2, & n+3, & \cdots, & \cdots, & 2n \\ \vdots & \vdots & \vdots & \vdots & \vdots & \vdots \\ (m-1)n+1, & (m-1)n+2, & (m-1)n+3, & \cdots, & \cdots, & mn \end{array}$$

We make the following observations about the above table (in what follows $1 \leq k, \ell \leq n$).

(i) If $gcd(k,n) > 1$ then $gcd(rn+k, n) > 1$. This is so because $gcd(k,n) = gcd(rn+k, n)$. Therefore if $gcd(k,n) > 1$ then $gcd(rn+k, mn) > 1$. So if $gcd(k,n) > 1$ then no number in the k-th column is co-prime to mn.

(ii) If $gcd(\ell, n) = 1$ then $gcd(rn + l, n) = 1$. So each of the number in the ℓ-th column is co-prime to n.

(iii) The m numbers in any column constitute a complete residue system modulo m. For if $0 \leq r, s < m$ and $rn+\ell \equiv sn+\ell (\text{mod } m)$ then $r \equiv s (\text{mod } m)$ and therefore $r = s$, since $0 \leq r, s < m$.

So among the mn numbers 1 through mn, the numbers which are co-prime to mn can appear only in columns which are headed by numbers which are co-prime to n - that is, these can appear in the ℓ-th columns where $gcd(\ell, n) = 1$. Since there are $\phi(n)$ such possible values for ℓ, there are exactly $\phi(n)$ columns which can contain numbers which are co-prime to mn. But, each of these columns contains m numbers which constitute a complete residue system modulo m, and therefore contains precisely $\phi(m)$ numbers which are co-prime to m. As these $\phi(m)$ numbers are already co-prime to n, in view of the fact that $gcd(r, mn) = gcd(r, m) \cdot gcd(r, n)$ (since $gcd(m, n) = 1$), it follows that there are $\phi(n)$ columns each contributing $\phi(m)$ numbers which are co-prime to mn. Therefore the numbers 1 through mn contains precisely $\phi(m) \cdot \phi(n)$ numbers which are co-prime to mn. Hence $\phi(mn) = \phi(m) \cdot \phi(n)$. This completes the proof of the theorem.

An Example:

We shall illustrate a few steps of the preceding proof by an example Let $m = 10,\ n = 9$. We write the numbers from 1 to 90 as array consisting of ten rows and nine columns.

$$\begin{array}{ccccccccc}
\boxed{\mathbf{1}} & \mathbf{2} & 3 & \mathbf{4} & \mathbf{5} & 6 & \boxed{\mathbf{7}} & \mathbf{8} & 9 \\
10 & \boxed{11} & 12 & \boxed{13} & 14 & 15 & 16 & \boxed{17} & 18 \\
\boxed{19} & 20 & 21 & 22 & \boxed{23} & 24 & 25 & 26 & 27 \\
28 & \boxed{29} & 30 & \boxed{31} & 32 & 33 & 34 & 35 & 36 \\
\boxed{37} & 38 & 39 & 40 & \boxed{41} & 42 & \boxed{43} & 44 & 45 \\
46 & \boxed{47} & 48 & \boxed{49} & 50 & 51 & 52 & \boxed{53} & 54 \\
55 & 56 & 57 & 58 & \boxed{59} & 60 & \boxed{61} & 62 & 63 \\
64 & 65 & 66 & \boxed{67} & 68 & 69 & 70 & \boxed{71} & 72 \\
\boxed{73} & 74 & 75 & 76 & \boxed{77} & 78 & \boxed{79} & 80 & 81 \\
82 & \boxed{83} & 84 & 85 & 86 & 87 & 88 & \boxed{89} & 90
\end{array}$$

In the first row we have bold faced $\phi(9) = 6$ entries which are co-prime to 9, namely $1, 2, 4, 5, 7$, and 8. The entries which contribute to $\phi(9 \times 10)$ can only appear in the columns headed by these six numbers. Each column consists of 10 numbers which constitute a complete residue system modulo 10 and therefore has exactly four entries (since $\phi(10) = 4$) which are co-prime to 10. So we have six columns each contributing four numbers each of which is co-prime to $9 \times 10 = 90$ giving us a total of $6 \times 4 = 24$ numbers in the array consisting of those numbers between 1 and 90 which are co-prime to 90. These are the the boxed numbers in the above array.

The above theorem namely $\phi(mn) = \phi(m) \cdot \phi(n)$ if $gcd(m, n) = 1$ can also be stated by saying that the Euler's ϕ-function is multiplicative. With the help of the Chinese Remainder Theorem we can give a different proof of the above theorem as follows. Let

$$\Omega_1 = \{x : 1 \leq x \leq m, gcd(x, m) = 1\}$$

$$\Omega_2 = \{y : 1 \leq y \leq n, gcd(y, n) = 1\}$$

$$\Omega = \{z : 1 \leq z \leq m, gcd(z, mn) = 1\}$$

Evidently $|\Omega_1| = \phi(m), |\Omega_2| = \phi(n)$ and $|\Omega| = \phi(mn)$. Consider the mapping

$$f : \Omega \longrightarrow \Omega_1 \times \Omega_2$$

defined as $f(z) = (z \bmod m,\ z \bmod n) = (z_1, z_2)$ where z_1 and z_2 are the least nonnegative residues mod m and mod n, respectively. Since $gcd(z, mn) = 1$, we have $gcd(z, m) = gcd(z, n) = 1$ and therefore $gcd(z_1, m) = 1$ and $gcd(z_2, n) = 1$. This mapping is easily seen to be one-to-one. By the Chinese Remainder Theorem the mapping f is onto. Therefore $\phi(mn) = \phi(m) \cdot \phi(n)$.

The rest of this section may be skipped by beginning students.

With the help of the Chinese Remainder Theorem we can give a proof of the following theorem in group theory which in turn can be used to study the structure of the group $\mathbb{Z}_n^*$ - the multiplicative group (modulo n) of integers less than n and co-prime to n (more precisely of the congruent classes modulo n, corresponding to the integers which are co-prime to n).

Theorem Let G be a cyclic group of order mn where $gcd(m, n) = 1$. Let H and K be its (unique) subgroups of order m and n respectively. Let $Aut(G)$, $Aut(H)$, and $Aut(K)$ (respectively) denote the groups of automorphism of the groups G, H, and K (respectively). Then the group $Aut(G)$ is isomorphic to the direct product of the groups $Aut(H)$ and $Aut(K)$, that is,

$$Aut(G) \approx Aut(H) \times Aut(K)$$

Proof Let $G = \langle z \rangle$. Then we can write $z = x \cdot y$ where x has order m and y has order n. Therefore we may assume that $H = \langle x \rangle$ and $K = \langle y \rangle$. Let σ be an automorphism of G. The restriction of σ to H and K (respectively) are automorphism of the subgroups H and K (respectively). The map

$$Aut(G) \longrightarrow Aut(H) \times Aut(K)$$

$$\sigma \longrightarrow (\sigma|H, \sigma|K)$$

is evidently a homomorphism of groups. Since the groups on the left and right hand side have same orders ($Aut(G)$ has order $\phi(mn)$ and $Aut(H) \times Aut(K)$ has order $\phi(m) \cdot \phi(n)$) it is enough to show that the above mapping is onto.

Let $\sigma_1 \in Aut(H)$, and $\sigma_2 \in Aut(K)$. Suppose that $\sigma_1(x) = x^r$ and $\sigma_2(y) = y^s$. Evidently $gcd(r, m) = gcd(s, n) = 1$. By Chinese Remainder Theorem we can find an integer t such that $t \equiv r (\text{mod } m)$ and $t \equiv s \ (\text{mod } n)$. Now $gcd(t, m) = gcd(r, m) = 1$ and $gcd(t, n) = gcd(s, n) = 1$. Therefore $gcd(t, mn) = 1$. If we let $\sigma \in Aut(G)$ be the mapping $\sigma(z) = z^t$, then $\sigma|H = \sigma_1$ and $\sigma|K = \sigma_2$. So the automorphism $\sigma \in Aut(G)$ is mapped to $(\sigma_1, \sigma_2) \in Aut(H) \times Aut(K)$. This proves that the above mapping is onto and therefore

$$Aut(G) \approx Aut(H) \times Aut(K)$$

This completes the proof of the theorem.

The above theorem can be easily generalized to the case when the cyclic group G is (isomorphic to) the direct product of more than two subgroups such that any two of these subgroups have co-prime orders. Using such a generalization of the above theorem we can determine the structure of the group $\mathbb{Z}_n^*$. It is clear that $\mathbb{Z}_n^* \approx Aut(C_n)$ where C_n denotes the cyclic group of order n. Let

$$n = p_1^{a_1} p_2^{a_2} \cdots p_k^{a_k}$$

where $p_1, p_2 \cdots p_k$ are distinct prime numbers. Then

$$C_n \approx C_{p_1^{a_1}} \times C_{p_2^{a_2}} \times \cdots C_{p_k^{a_k}}$$

Therefore, using above theorem we have

$$Aut(C_n) \approx Aut(C_{p_1^{a_1}}) \times Aut(C_{p_2^{a_2}}) \times \cdots Aut(C_{p_k^{a_k}})$$

That is

$$\mathbb{Z}_n^* \approx \mathbb{Z}_{p_1^{a_1}}^* \times \mathbb{Z}_{p_2^{a_2}}^* \times \cdots \mathbb{Z}_{p_k^{a_k}}^*$$

In other words once we know the structure of the group $\mathbb{Z}_n^*$ for prime power value of n then we know the structure of the group $\mathbb{Z}_n^*$ for all values of n. We shall determine the structure of the groups $\mathbb{Z}_n^*$ when n is a prime power in the next chapter.

2. Greatest Integer Function

For any real number x the greatest integer function $[x]$ is defined to be the largest integer less than or equal to x. For example $[2.7] = 2$, $[-5.3] = -6$, $[\pi] = 3$, $[4] = 4$. If x and y are real numbers then it can be easily seen that $[x + y] \geq [x] + [y]$. If n and k are positive integers then by the division algorithm we can find integers q and r such that $n = k \cdot q + r$, where $0 \leq r < k$. Then $\dfrac{n}{q} = k + \dfrac{r}{q}$, $0 \leq \dfrac{r}{q} < 1$. Therefore $\left[\dfrac{n}{q}\right] = k$, is the quotient when n is divided by q. The following lemma is an immediate consequence of the above observation.

Lemma Let n and k be positive integers. The number of integers between 1 and n (inclusive) which are divisible by k is precisely $\left[\dfrac{n}{k}\right]$.

The following theorem (known as de Polignac's formula) gives information about the largest exponent of a prime p which divides $n!$.

Theorem Let n be a positive integer and p be a prime number. Then the largest exponent e such that p^e divides $n!$ is

$$e = \sum_{i=1}^{\infty} \left[\frac{n}{p^i}\right]$$

Proof For a positive integer k, let $S_n(k)$ denote the number of integers between 1 and n (inclusive) which are divisible by k. Obviously $e = \sum_{i=1}^{\infty} S_n(p^i)$. From the lemma, $S_n(p^i) = \left[\frac{n}{p^i}\right]$. Hence

$$e = \sum_{i=1}^{\infty} \left[\frac{n}{p^i}\right]$$

Remark With the help of the preceding theorem we can compute the number of trailing zeros when $n!$ is written in decimal notation. Evidently the number of trailing zeros in $n!$ is the exact power of 10 that divides n factorial. Since there is no shortage of powers of 2 in n factorial (that is in comparison with powers of 5) all that we need to compute is the largest e such that 5^e divides n factorial. For example the number of trailing zeros in 365! is given by the expression

$$\sum_{i=1}^{\infty} \left[\frac{365}{5^i}\right] = \left[\frac{365}{5}\right] + \left[\frac{365}{5^2}\right] + \left[\frac{365}{5^3}\right] = 73 + 14 + 2 = 89$$

One can easily check the truth of the preceding statement with the help of suitable software such as Maple or Mathematica.

3. The functions $\tau(n)$, $\sigma(n)$, and $\mu(n)$

These three arithmetical functions are defined as follows:

- The number divisors of n, $\tau(n) = \sum_{d|n} 1$
- The sum of divisors of n, $\sigma(n) = \sum_{d|n} d$
- The Mobius μ-function is defined as

$$\mu(n) = \begin{cases} 0 & \text{if } n \text{ is divisible by some square} > 1 \\ (-1)^k & \text{if } n = p_1 \cdot p_2 \cdots p_k \text{ is a product of } k \text{ distinct primes} \end{cases}$$

Each of the above mentioned arithmetical functions is a multiplicative (but not completely multiplicative) function. We have already proved that the Euler's ϕ-function is a multiplicative function. We shall use the following theorem to prove that the functions τ and σ are multiplicative arithmetical functions.

Theorem Let f be a multiplicative arithmetical function. Then the arithmetical function g defined by

$$g(n) = \sum_{d|n} f(d)$$

is also multiplicative.

Proof Let m and n be positive integers such that $gcd(m, n) = 1$. Then

$$\begin{aligned} g(mn) &= \textstyle\sum_{d|mn} f(d) = \sum_{d_1|m, d_2|n} f(d_1 d_2) = \sum_{d_1|m, d_2|n} f(d_1) f(d_2) \\ &= \textstyle\sum_{d_2|n} f(d_2) \sum_{d_1|m} f(d_1) = \sum_{d_2|n} f(d_2) \cdot g(m) = g(m) \sum_{d_2|n} f(d_2) = g(m) \cdot g(n) \end{aligned}$$

Corollary The functions τ and σ are multiplicative.

Proof Follows immediately from the preceding theorems - by taking $f(d) = 1$ in the case of function τ and $f(d) = d$ in the case of function σ.

Theorem The Mobius μ-function is multiplicative.

Proof Let m and n be two positive integers such that $gcd(m, n) = 1$. Evidently mn is divisible by some square (greater than one) if and only if one of the numbers m or n is divisible by some square. This shows that $\mu(mn) = 0$ if and only if either $\mu(m)$ or $\mu(n)$ is zero. Now suppose that both m and n are square-free, that is neither m nor n is divisible by square of any integer $k > 1$. So both m and n are products of distinct primes. Let $m = p_1 \cdot p_2 \cdots p_r$ and $n = q_1 \cdot q_2 \cdots q_s$ where $p_1, p_2, \cdots, p_r$ and $q_1, q_2, \cdots, q_s$ are distinct prime numbers. Then $\mu(m) = (-1)^r$ and $\mu(n) = (-1)^s$ and $\mu(mn) = \mu(p_1, p_2, \cdots, p_r q_1, q_2, \cdots, q_s) = (-1)^{r+s}$. So in every case we have $\mu(mn) = \mu(m)\mu(n)$. This proves that μ is a multiplicative function.

Computation of $\phi(n)$, $\tau(n)$, and $\sigma(n)$.

If p is a prime number then as we have seen earlier $\phi(p^r) = p^r - p^{r-1}$. Since the divisors of p^r are $1, p, p^2, \cdots, p^r$, we have $\tau(p^r) = r + 1$ and $\sigma(p^r) = 1 + p + p^2 + \cdots + p^r = \dfrac{p^{r+1} - 1}{p - 1}$. Let $n = p_1^{a_1} \cdot p_2^{a_2} \cdots p_k^{a_k}$, where p_i, $1 \leq i \leq k$ are distinct primes. Then

$$\phi(n) = \prod_{i=1}^{k} \phi(p_i^{a_i}) = \prod_{i=1}^{k} (p_i^{a_i} - p_i^{a_i - 1}) = \prod_{i=1}^{k} p_i^{a_i} \left(1 - \frac{1}{p_i}\right) = n \prod_{i=1}^{k} \left(1 - \frac{1}{p_i}\right)$$

$$\tau(n) = \prod_{i=1}^{k} \tau(p_i^{a_i}) = \prod_{i=1}^{k} (a_i + 1)$$

$$\sigma(n) = \prod_{i=1}^{k} \sigma(p_i^{a_i}) = \prod_{i=1}^{k} \left(\frac{p_i^{a_i+1} - 1}{p_i - 1}\right)$$

4. Perfect Numbers

A positive integer n is said to be a *perfect number* if $\sigma(n) = 2n$. Perfect numbers have been investigated and of interest since antiquity. The first three perfect numbers are 6, 28, and 496. All perfect numbers known thus far are even. It is conjectured that every perfect number is even. It is also conjectured that there are infinitely many of them. Neither of these conjectures have been proven yet. However, we have the following theorem giving the description of even perfect numbers.

Theorem Let N be an even positive integer. Then N is a perfect number if and only if $N = 2^{p-1}(2^p - 1)$ where p is a prime number such that $q = 2^p - 1$ is also a prime number.

Proof First suppose that $N = 2^{p-1} \cdot q$, where $q = 2^p - 1$ is a prime number. Then

$$\sigma(N) = \sigma(2^{p-1} \cdot q) = \sigma(2^{p-1}) \cdot \sigma(q) = (2^p - 1)(q + 1) = q \cdot 2^p = 2N.$$

So N is a perfect number. Now suppose that N is an even perfect number. We can write $N = 2^k \cdot q$ where q is odd and $k \geq 1$. Since N is given to be a perfect number we have $\sigma(N) = 2N$. We note that $p > 1$ is a prime number if and only if $\sigma(p) = p + 1$. It is clear that if p is a prime number then $\sigma(p) = p+1$. On the other hand if p is a composite number then it has a divisor other than 1 and p. Therefore $\sigma(p) > p + 1$. Since $N = 2^k \cdot q$ (q odd), we have

$$\sigma(N) = \sigma(2^k) \cdot \sigma(q) = (2^{k+1} - 1)\sigma(q)$$

Since N is a perfect number $\sigma(N) = 2N = 2^{k+1}q$. Therefore

$$(2^{k+1} - 1)\sigma(q) = 2^{k+1} \cdot q.$$

Since q is odd, it follows that 2^{k+1} divides $\sigma(q)$. Let $\sigma(q) = 2^{k+1} \cdot r$ for some r. Substituting this value of $\sigma(q)$ in the above relation we get $q = (2^{k+1} - 1)r$ and so we have

$$\sigma(q) = 2^{k+1}r = (2^{k+1} - 1)r + r = q + r.$$

If $r > 1$ then q has divisors (among possibly others) $1, r$, and q. Therefore $\sigma(q) \geq q + r + 1$. This contradicts the fact that $\sigma(q) = q + r$. So we must have $r = 1$ and $\sigma(q) = 2^{k+1} = q + 1$. It follows that q is a prime number. Now as $q = 2^{k+1} - 1$ is a prime number therefore $k + 1$ is a prime number. Let $k + 1 = p$ and so $q = 2^p - 1$. Then

$$N = 2^k \cdot q = 2^{p-1}(2^p - 1)$$

where, $q = 2^p - 1$ and p are prime numbers. This completes the proof of the theorem.

Note The primes of the form $2^p - 1$ are known as Mersenne Primes. Appendix 1 gives a list of several Mersenne prime numbers.

We showed earlier that Euler's ϕ-function is multiplicative. Now we shall state and prove two theorems concerning Euler's ϕ-function.

Theorem Let n be a positive integer. Then

$$\sum_{d|n} \phi(d) = n$$

First Proof

Let

$$g(n) = \sum_{d|n} \phi(d)$$

Since ϕ is a multiplicative function, so is the function g. So it is sufficient to prove that the theorem when $n = p^k$, a power of a prime number p. We have

$$g(p^k) = \sum_{d|p^k} \phi(d) = \phi(1)+\phi(p)+\phi(p^2)+\cdots+\phi(p^k) = 1+(p-1)+(p^2-p)+\cdots+(p^k-p^{k-1}) = p^k$$

So $g(p^k) = p^k$. Since g is a multiplicative function, it follows that $g(n) = n$ for all n which completes the proof of the theorem.

Second Proof

For the second proof we shall use the following well-known results from group theory.

(i) If G is a cyclic group of order n then for each divisor d of n it has a unique (cyclic) subgroup of order d.

(ii) A cyclic group of order n has exactly $\phi(n)$ elements of order n.

Let G be cyclic group of order n. Let $x \in G$ be an element of order d. Then by Lagrange's Theorem in group theory d divides n and x belongs to the unique cyclic subgroup of G of order d. For each d dividing n let H_d denote this unique cyclic subgroup of G of order d. Then counting the elements of G in two different ways we have:

$|G| = \sum_{d|n} \#\{\text{Elements of order } d \text{ in } G\} = \sum_{d|n} \#\{\text{generators of } H_d\} = \sum_{d|n} \phi(d)$

That is $n = \sum_{d|n} \phi(d)$.

Theorem Let $n > 1$ be a positive integer. Then

$$\sum_{1\leq d\leq n,(d,n)=1} d = \frac{1}{2}\phi(n)$$

Proof Clearly, $gcd(d,n) = 1$ if and only if $gcd(n-d,n) = 1$. Therefore

$$\sum_{1\leq d\leq n,(d,n)=1} d = \sum_{1\leq d\leq n,(d,n)=1} (n-d) = \sum_{1\leq d\leq n,(d,n)=1} n - \sum_{1\leq d\leq n,(d,n)=1} d = n\cdot\phi(n) - \sum_{1\leq d\leq n,(d,n)=1} d$$

Hence

$$\sum_{1\leq d\leq n,(d,n)=1} d = \frac{1}{2}n\phi(n)$$

5. Mobius Inversion Formula

We shall now prove the Mobius Inversion Formula which is very useful in number theory and also give a few of its applications.

Theorem Let f be an arithmetical function and let

$$F(n) = \sum_{d|n} f(d)$$

then

$$f(n) = \sum_{d|n} F(d)\mu(\frac{n}{d})$$

Proof

$$\begin{aligned}
\textstyle\sum_{d|n} F(d)\mu(\frac{n}{d}) &= \textstyle\sum_{d|n} \mu(d)\cdot F(\frac{n}{d}) \\
&= \textstyle\sum_{d|n} \mu(d)\cdot \sum_{c|\frac{n}{d}} f(c) \\
&= \textstyle\sum_{cd|n} \mu(d)\cdot f(c) \\
&= \textstyle\sum_{cd|n} f(d)\mu(c) \\
&= \textstyle\sum_{d|n} f(d)\cdot \sum_{c|\frac{n}{d}} \mu(c) \\
&= f(n)
\end{aligned}$$

The last equality follows from the observation that

$$\sum_{c|\frac{n}{d}} \mu(c) = \begin{cases} 1 & \text{if } \frac{n}{d} = 1 \\ 0 & \text{otherwise} \end{cases}$$

Hence

$$f(n) = \sum_{d|n} F(d)\mu\left(\frac{n}{d}\right)$$

The converse of the above theorem is true and is often used.

Theorem If F is an arithmetical function and the function f is defined by

$$f(n) = \sum_{d|n} \mu(d)F\left(\frac{n}{d}\right)$$

then

$$F(n) = \sum_{d|n} f(d)$$

Proof We have

$$\begin{aligned}
\textstyle\sum_{d|n} f(d) &= \textstyle\sum_{d|n}\sum_{r|d}\mu(r)F(\frac{d}{r}) \\
&= \textstyle\sum_{rs|n}\mu(r)F(s) \\
&= \textstyle\sum_{s|n}\sum_{r|\frac{n}{s}}\mu(r)F(s) \\
&= \textstyle\sum_{s|n}F(s)\sum_{r|\frac{n}{s}}\mu(r) \\
&= F(n)
\end{aligned}$$

Since

$$\sum_{r|\frac{n}{s}} \mu(i) = \begin{cases} 1 & \text{if } \dfrac{n}{s} = 1 \\ 0 & \text{otherwise} \end{cases}$$

The converse of the Mobius Inversion Formula can be used to give yet another proof of the fact that $\sum_{d|n} \phi(d) = n$. In order to give this proof we need the following lemma.

Lemma If $n \in \mathbb{N}$ then

$$\sum_{d|n} \frac{\mu(d)}{d} = \prod_{p|n}\left(1 - \frac{1}{p}\right)$$

where the last product is over all prime divisors p of n.

Proof Let

$$g(n) = \sum_{d|n} \frac{\mu(d)}{d}$$

The function g is easily seen to be multiplicative. So it is enough to prove that for a prime power p^r, $g(p^r) = \left(1 - \dfrac{1}{p}\right)$. Now

$$g(p^r) = \sum_{d|p^r} \frac{\mu(d)}{d} = 1 - \frac{1}{p}$$

Now we give another proof of $\sum_{d|n} \phi(d) = n$ using the converse of the Möbius Inversion Formula. Let F be the identity function, that is, $F(n) = n$ for all n. Let

$$f(n) = \sum_{d|n} \mu(d) F\left(\frac{n}{d}\right)$$

Then

$$f(n) = \sum_{d|n} \mu(d) \cdot \frac{n}{d} = n \cdot \sum_{d|n} \frac{\mu(d)}{d} = n \cdot \prod_{p|n} \left(1 - \frac{1}{p}\right) = \phi(n)$$

where the last product is over all the prime divisors p of n. So the function f is nothing but the Euler's ϕ-function. By the converse of the Mobius Inversion Formula we have

$$F(n) = \sum_{d|n} f(d)$$

or

$$n = \sum_{d|n} \phi(d)$$

Multiplicative Form of Mobius Inversion Formula

The multiplicative form of Mobius Inversion Formula states that if $f(n)$ is an arithmetical function and the function F is defined as

$$F(n) = \prod_{d|n} f(d)$$

then

$$f(n) = \prod_{d|n} F(d)^{\mu(\frac{n}{d})}$$

We shall give two proofs of this.

First Proof

$$\begin{aligned}
\prod_{d|n} F(d)^{\mu(\frac{n}{d})} &= \prod_{d|n} F(\tfrac{n}{d})^{\mu(d)} \\
&= \prod_{d|n} \left(\prod_{c|\frac{n}{d}} f(c)\right)^{\mu(d)} \\
&= \prod_{cd|n} f(c)^{\mu(d)} \\
&= \prod_{cd|n} f(d)^{\mu(c)} \\
&= \prod_{d|n} \prod_{c|\frac{n}{d}} f(d)^{\mu(c)} \\
&= \prod_{d|n} f(d)^{\sum_{c|\frac{n}{d}} \mu(c)} \\
&= f(n)
\end{aligned}$$

Since $\sum_{c|\frac{n}{d}} \mu(c) = \begin{cases} 1 \text{ if } \frac{n}{d} = 1 \\ 0 \text{ otherwise} \end{cases}$

Second Proof We have $F(n) = \prod_{d|n} f(d)$ Taking logs of both sides we get

$$\log(F(n)) = \sum_{d|n} \log(f(d))$$

By Mobius Inversion Formula we have

$$\begin{aligned} \log(f(n)) &= \textstyle\sum_{d|n} \mu(\frac{n}{d}) \log F(d) \\ &= \textstyle\sum_{d|n} \mu(d) \log F(\frac{n}{d}) \\ &= \textstyle\sum_{d|n} \log(F(\frac{n}{d}))^{\mu(d)} \\ &= \log\left(\textstyle\prod_{d|n} F(\frac{n}{d})^{\mu(d)}\right) \end{aligned}$$

Therefore

$$f(n) = \prod_{d|n} F\left(\frac{n}{d}\right)^{\mu(d)}$$

Application of Mobius Formula in Computing Cyclotomic Polynomials

We recall the n-th cyclotomic polynomial $\Phi_n(x)$ is defined as:

$$\Phi_n(x) = \prod_{gcd(r,n)=1} (x - e^{\frac{2\pi i r}{n}})$$

These polynomials have the following properties

- $\Phi_n(x) \in \mathbb{Z}[x]$ is monic and irreducible.
- $\Phi_n(x)$ has degree $\phi(n)$.

We have

$$x^n - 1 = \prod_{d|n} \Phi_d(x)$$

By applying the multiplicative version of the Mobius formula we have

$$\Phi_n(x) = \prod_{d|n} (x^d - 1)^{\mu(\frac{n}{d})}$$

Two Examples

Example 1. Let $n = 12$. Using the above formula we have

$$\begin{aligned} \Phi_{12}(x) &= \textstyle\prod_{d|12} (x^d - 1)^{\mu(\frac{12}{d})} \\ &= (x-1)^{\mu(12)} \cdot (x^2-1)^{\mu(6)} \cdot (x^3-1)^{\mu(4)} \cdot (x^4-1)^{\mu(3)} \cdot (x^6-1)^{\mu(2)} \cdot (x^{12}-1)^{\mu(1)} \\ &= \cdot(x^2-1)^{\mu(6)} \cdot (x^4-1)^{\mu(3)} \cdot (x^6-1)^{\mu(2)} \cdot (x^{12}-1)^{\mu(1)} \\ &= (x^2-1) \cdot \frac{1}{x^4-1} \cdot \frac{1}{x^6-1} \cdot (x^{12}-1) \\ &= \frac{x^6+1}{x^2+1} = x^4 - x^2 + 1. \end{aligned}$$

Example 2. Let $n = 20$. Using the above formula we have

$$\begin{aligned}
\Phi_{20}(x) &= \prod_{d|20}(x^d - 1)^{\mu(\frac{20}{d})} \\
&= (x-1)^{\mu(20)} \cdot (x^2-1)^{\mu(10)} \cdot (x^4-1)^{\mu(5)} \cdot (x^5-1)^{\mu(4)} \cdot (x^{10}-1)^{\mu(2)} \cdot (x^{20}-1)^{\mu(1)} \\
&= \cdot(x^2-1)^{\mu(10)} \cdot (x^4-1)^{\mu(5)} \cdot (x^{10}-1)^{\mu(2)} \cdot (x^{20}-1)^{\mu(1)} \\
&= (x^2-1) \cdot \frac{1}{x^4-1} \cdot \frac{1}{x^{10}-1} \cdot (x^{20}-1) \\
&= \frac{x^{10}+1}{x^2+1} = x^8 - x^6 + x^4 - x^2 + 1.
\end{aligned}$$

Example 3. Let $n = 100$. Since the only square-free divisors of 100 are $1, 2, 5$, and 10, using above formula we have

$$\begin{aligned}
\Phi_{100}(x) &= \prod_{d|100}(x^d - 1)^{\mu(\frac{100}{d})} \\
&= (x^{100}-1)^{\mu(1)} \cdot (x^{50}-1)^{\mu(2)} \cdot (x^{20}-1)^{\mu(5)} \cdot (x^{10}-1)^{\mu(10)} \\
&= \frac{x^{50}+1}{x^{10}+1} = x^{40} - x^{30} + x^{20} - x^{10} + 1.
\end{aligned}$$

6. Counting The Number of Monic Irreducible Polynomials Over a Finite Field

In this section we shall give a useful application of the Möbius Inversion Formula. We shall deduce a formula giving the number of distinct monic irreducible polynomials of degree n over a finite field. This section requires some knowledge of finite fields and may be omitted at first reading.

It is known that a finite field of order q exists if and only if q is a power of a prime number. Let $\mathbb{K} = GF(q)$ be a finite field of q elements, where q is a power of a prime number. We first prove the following lemma.

Lemma Let $f(x) \in \mathbb{K}[x]$ be an irreducible polynomial of degree d and n be positive integer. Then d divides n if and only if $f(x)$ divides $x^{q^n} - x$.

Proof First suppose that $f(x)$ divides $x^{q^n} - x$. Let $x^{q^n} - x = f(x) \cdot g(x)$, for some polynomial $g(x) \in \mathbb{K}[x]$. Let $\mathbb{F} = GF(q^d)$ denote the splitting field of $f(x)$. Without loss of generality we may assume that $\mathbb{F} = GF(q^d)$ is contained in $GF(q^n)$, the splitting field of $x^{q^n} - x$ over $\mathbb{K} = GF(q)$. Consider the following diagram of fields

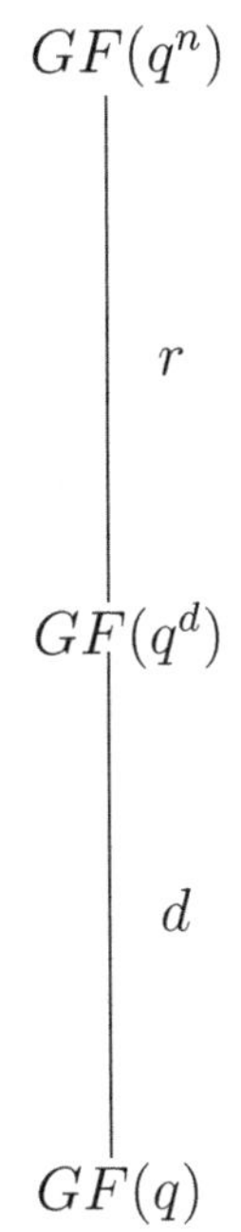

Let $GF(q^n)$ be of dimension r (as a vector space) over $GF(q^d)$. Then $n = r \cdot d$ and so d divides n. Conversely, suppose that d divides n. It follows that $q^d - 1$ divides $q^n - 1$ and therefore $x^{q^d-1} - 1$ divides $x^{q^n-1} - 1$. Since every element of $GF(q^d)$ satisfies the polynomial equation $x^{q^d} - x = 0$, it follows that $f(x)$ divides $x^{q^d} - x$. Now $f(x)$ divides $x^{q^d} - x$ which in turn divides $x^{q^n} - x$. Hence $f(x)$ divides $x^{q^n} - x$.

Theorem Let $\mathbb{K} = GF(q)$ be a finite field of q elements and $\Psi(n)$ denote the number of monic irreducible polynomials of degree n over K. Then

$$\Psi(n) = \frac{1}{n} \sum_{d|n} q^{\frac{n}{d}} \mu(d)$$

Proof Let as before $GF(q^n)$ denote a finite field of q^n elements containing $GF(q)$. We can think of $GF(q^n)$ as the splitting field of $x^{q^n} - x$ over $\mathbb{K}$. If we write $x^{q^n} - x$ as a product of (monic) irreducible polynomials over $\mathbb{K}$, then the factors must be distinct - since the elements of $GF(q^n)$ are precisely the q^n distinct zeros of $x^{q^n} - x$, and their degrees must divide n as proved in the previous lemma. Therefore

$$x^{q^n} - x = \prod_{d|n} \left(\prod_{d} f_d(x) \right)$$

where $\prod_d f_d(x)$ denotes the product of all monic irreducible polynomials of degree d over $GF(q)$. Counting degrees on both sides we have

$$q^n = \sum_{d|n} d \cdot \Psi(d)$$

where (we recall) $\Psi(d)$ denotes the number of monic irreducible polynomials of degree d over $GF(q)$. Using Möbius Inversion Formula we get

$$n\Psi(n) = \sum_{d|n} q^{\frac{n}{d}} \mu(d)$$

or

$$\Psi(n) = \frac{1}{n} \sum_{d|n} q^{\frac{n}{d}} \mu(d)$$

This completes the proof of the theorem. The following table gives the values of $\Psi(n)$ for various values of n and q.

n	$\Psi(n)$	The value of $\Psi(n)$ for	
		$q = 2$	$q = 3$
1	q	2	3
2	$\frac{1}{2}(q^2 - q)$	1	3
3	$\frac{1}{3}(q^3 - q)$	2	8
4	$\frac{1}{4}(q^4 - q^2)$	3	18
5	$\frac{1}{5}(q^5 - q)$	6	80
6	$\frac{1}{6}(q^6 - q^3 - q^2 + q)$	9	16

In particular the eight monic irreducible polynomials of degree three $GF(3)$ are:

(i) $x^3 + x^2 + 2$

(ii) $x^3 + 2x^2 + 1$

(iii) $x^3 + 2x + 2$

(iv) $x^3 + 2x + 1$

(v) $x^3 + x^2 + x + 2$

(vi) $x^3 + x^2 + 2x + 1$

(vii) $x^3 + 2x^2 + x + 1$

(viii) $x^3 + 2x^2 + 2x + 2$

CHAPTER 3

Primitive Roots and the Structure of Groups $\mathbb{Z}_n^*$

1. The Group $\mathbb{Z}_n^*$

For $n \geq 1$ the set of congruence classes modulo n forms a ring denoted by $\mathbb{Z}_n$. The set of elements of $\mathbb{Z}_n$ which have multiplicative inverses form a (multiplicative) group denoted by $\mathbb{Z}_n^*$. As an element $\bar{x} \in \mathbb{Z}_n$ has a multiplicative inverse if and only if $gcd(x, n) = 1$. Therefore, the order of the group $\mathbb{Z}_n^*$ is $\phi(n)$. We shall show that only values of n for which the group $\mathbb{Z}_n^*$ is cyclic are $n = 1, 2, 4, p^k, 2p^k$ where p is an odd prime. A positive integer n is said to have a *primitive root* if $\mathbb{Z}_n^*$ is a cyclic group. So $n \in \mathbb{N}$ has a primitive root if and only if $n = 1, 2, 4, p^k, 2p^k$ where p is an odd prime. If n has a primitive root then an integer x such that $\bar{x}$ generates $\mathbb{Z}_n^*$ is called a primitive root modulo n. From group theory we know that a cyclic group of order m has precisely $\phi(m)$ elements of order m. Therefore, if n has a primitive root then it has precisely $\phi(\phi(n))$ primitive roots. The main purpose of this chapter is to prove that if the natural number n has a primitive root then $n = 1, 2, 4, p^k$ or $2p^k$ where p is an odd prime. In order to be able to do this we shall need some intermediate results which we now discuss.

Theorem A finite subgroup of the multiplicative group of a field is cyclic.

Proof Let $\mathbb{F}$ be a field and $G \subseteq \mathbb{F}^* = \mathbb{F} \setminus \{0\}$ be a finite group of order n. Recall that the exponent of a finite group is the least common multiple of the order of its elements. Let t be the exponent of G. As G is abelian it has an element of order t. Consider the polynomial $x^t - 1 \in \mathbb{F}[x]$. Every element of G is a zero of the polynomial $x^t - 1$. So the polynomial $x^t - 1$ of degree t has at least n zeros. If $t < n$ this is not possible. So we must have $t = n$ and G has an element of order n. Thus G is a cyclic group.

2. Fermat's Little Theorem

Theorem: (Fermat) If p is any prime and a is any integer such that $(a, p) = 1$ then $a^{p-1} \equiv 1 \pmod{p}$.

Proof: It is an immediate consequence of the fact that $\mathbb{Z}_p^*$ is a (multiplicative) group of the order p-1.

The above theorem can also be stated as $a^p \equiv a (\text{mod } p)$ for all a. It is interesting to observe that it can also be proved by induction since $(a+1)^p \equiv a^p + 1 \equiv a + 1 (\text{mod } p)$.

3. Euler's Generalization of the Fermat's Little Theorem

Let n be any positive integer and a any integer such that $gcd(a, n) = 1$. Since $\mathbb{Z}_n^*$ is a group (with the binary operation of multiplication modulo n) of order $\phi(n)$, Lagrange's theorem (namely that order of a subgroup of a finite group divides the order of the group) implies that $a^{\phi(n)} \equiv 1 (\text{mod } n)$. This statement is known as Euler's generalization of Fermat's Little Theorem. If $n = p$ a prime number then $\phi(p) = p - 1$, the above statement becomes $a^{p-1} \equiv 1$ (mod p) - which justifies it being termed as a generalization of Fermat's Little Theorem.

The following theorem gives us the structure of the group $\mathbb{Z}_n^*$ when n is a power of a prime number.

Theorem If p is an odd prime then $\mathbb{Z}_{p^n}^* \simeq C_{p^{n-1}(p-1)}$. $\mathbb{Z}_2^* \simeq C_1$, $\mathbb{Z}_4^* \simeq C_2$ and $\mathbb{Z}_{2^n}^* \simeq C_{2^{n-2}} \times C_2$ for $n > 2$.

In other word, the automorphism group of a cyclic group of order p^n (p an odd prime) is itself cyclic of order $p^{n-1}(p-1)$. The automorphism group of a cyclic group of order 2^n ($n > 2$) is a direct product of a cyclic group of order 2^{n-2} and a cyclic group of order 2.

We shall need the following lemma for the proof of the above theorem.

Lemma

(1) Let p be an odd prime and $k \geq 1$. If x is any integer such that $x \equiv 1$ (mod p^k), but $x \not\equiv 1 (\text{mod } p^{k+1})$ then $x^p \equiv 1 (\text{mod } p^{k+1})$, and $x^p \not\equiv 1 (\text{mod } p^{k+2})$.

(2) The same result holds if $p = 2$, $k \geq 2$. (The only exception is case is $p = 2, k = 1$).

Proof

(1) Let $x = 1 + tp^k$, $p \not| \; t$. So $x \not\equiv 1$ (mod p^{k+1}). We have

$$\begin{aligned} x^p &= (1 + tp^k)^p \\ &= 1 + \binom{p}{1} tp^k + \binom{p}{2} (tp^k)^2 + \cdots + \binom{p}{p-1} (tp^k)^{p-1} + (tp^k)^p \\ &\equiv 1 \quad (\text{mod } p^{k+1}) \end{aligned}$$

Since each of the binomial coefficients $\binom{p}{i}$, $1 \le i \le p-1$ is a multiple of p it follows that (since $p > 2$) every term following the term $\binom{p}{1}tp^k$ is a multiple of p^{k+2}. Since $\binom{p}{1}tp^k \equiv 0 (\text{mod } p^{k+1})$, but $\binom{p}{1}tp^k \not\equiv 0(\text{mod } p^{k+2})$, it follows that $x^p \not\equiv 1(\text{mod } p^{k+2})$.

(2) Let $p = 2$, $k \ge 2$. In this case we can write $x = 1 + t2^k$, where t is odd. We have $x^2 = 1 + t \cdot 2^{k+1} + t^2 2^{2k}$. Therefore, as before, $x^2 \equiv 1 \pmod{2^{k+1}}$ but $x^2 \not\equiv 1(\text{mod } 2^{k+2})$.

As mentioned earlier, the conclusion is not valid if $p = 2$ and $k = 1$. In fact if we let $x = 3$, then $3 \equiv 1(\text{mod } 2)$, $3 \not\equiv 1 \pmod{2^2}$, but $3^2 \equiv 1(\text{mod } 2^3)$.

Proof of the theorem

Part I. Let p be an odd prime. Consider the mapping

$$\begin{array}{ccc} \mathbb{Z}_{p^n}^* & \longrightarrow & \mathbb{Z}_p^* \\ x & \longrightarrow & x \pmod p \end{array}$$

This mapping is evidently an epimorphism. Under a homomorphism the order of the image of x divides the order of x. Since $\mathbb{Z}_p^*$ is a cyclic group of order $p-1$, we can find an element c in $\mathbb{Z}_{p^n}^*$ such that its order is a multiple of $p-1$. So a suitable power of c has order $p-1$. Thus $\mathbb{Z}_{p^n}^*$ has an element of order $p-1$. Let $x = p+1$. Then $x \equiv 1(\text{mod } p)$, $x \not\equiv 1 \pmod{p^2}$. By repeated application of the lemma we get $x^{p^{n-1}} \equiv 1(\text{mod } p^n)$, $x^{p^{n-2}} \not\equiv 1(\text{mod } p^n)$, So $x = p+1$ is an element of order p^{n-1} in $\mathbb{Z}_{p^n}^*$. Thus we have shown that the Abelian group $\mathbb{Z}_{p^n}^*$ has elements of order $p-1$ and of order p^{n-1}. Since $p-1$ and p^{n-1} are relatively prime it follows that $\mathbb{Z}_{p^n}^*$ has an element of order $p^{n-1}(p-1)$—the order of the group $\mathbb{Z}_{p^n}^*$. Hence the group $\mathbb{Z}_{p^n}^*$ is a cyclic group.

Remark Very often a generator of $\mathbb{Z}_p^*$ serves as a generator of $\mathbb{Z}_{p^n}^*$ for $n > 1$. This is not true in general. A counter example to this situation is seen for $p = 487$. In this case, $x = 10$ has order 486 in $\mathbb{Z}_{487}^*$ as well as in $\mathbb{Z}_{487^2}^*$. So $x = 10$ generates $\mathbb{Z}_{487}^*$ but not $\mathbb{Z}_{487^2}^*$

Part II. We now deal with the case $p = 2$. Since $\mathbb{Z}_2^*$ and $\mathbb{Z}_4^*$ are cyclic groups we only need to prove that for $n > 2$, $\mathbb{Z}_{2^n}^* \simeq C_2 \times C_{2^{n-2}}$. Let $x = 5$. So $x \equiv 1(\text{mod } 2^2)$, $x \not\equiv 1(\text{mod } 2^3)$. By the Lemma, $x^{2^{n-2}} \equiv 1(\text{mod } 2^n)$, $x^{2^{n-3}} \not\equiv 1(\text{mod } 2^n)$. So $x = 5$ is an element of order 2^{n-2} in $\mathbb{Z}_{2^n}^*$. Evidently -1 is an element of order 2 in $\mathbb{Z}_{2^n}^*$. Since $-1 \equiv 3(\text{mod } 4)$ and every element of the subgroup $\langle 5 \rangle$ generated by 5 in $\mathbb{Z}_{2^n}^*$ is congruent to $1(\text{mod } 4)$, it follows that -1 does not belong to the subgroup generated by 5. Therefore $\langle 5 \rangle \cdot \langle -1 \rangle \simeq \langle 5 \rangle \times \langle -1 \rangle$ is a subgroup of $\mathbb{Z}_{2^n}^*$ of order $2^{n-2} \cdot 2 = 2^{n-1} = \phi(2^n)$, the order of the group $\mathbb{Z}_{2^n}^*$. Hence $\mathbb{Z}_{2^n}^* \simeq C_{2^{n-2}} \times C_2$.

We now prove the following result which gives a characterization of integers having a primitive root.

Theorem The group $\mathbb{Z}_N^*$ is cyclic only for the following values $N = 1, 2, 4, p^n, 2p^n$; where p is an odd prime. Equivalently these are the only values of N for which N has a primitive root.

Proof We have seen that $\mathbb{Z}_N^*$ is cyclic if $N = 1, 2, 4$ or p^n, where p is an odd prime. Let $N = 2 \cdot p^n$. Then

$$\mathbb{Z}_{2\cdot p^n}^* \simeq \mathbb{Z}_2^* \times \mathbb{Z}_{p^n}^* \simeq \mathbb{Z}_{p^n}^*$$

Since $\mathbb{Z}_{p^n}^*$ is cyclic it follows that $\mathbb{Z}_{2\cdot p^n}^*$ is cyclic. Now suppose that $\mathbb{Z}_N^*$ is cyclic. We shall show that $N = 1, 2, 4, p^n$ or $2 \cdot p^n$, where p is an odd prime. Let $N = p_1^{a_1} \cdot p_2^{a_2} \cdots p_k^{a_k}$, where $p_i's$ are distinct primes. Then

$$\mathbb{Z}_N^* \simeq \mathbb{Z}_{p_1^{a_1}}^* \times \mathbb{Z}_{p_2^{a_2}}^* \times \cdots \times \mathbb{Z}_{p_k^{a_k}}^*$$

We note that the order of $\mathbb{Z}_{p_i^{a_i}}^*$ is $\phi(p_i^{a_i}) = p_i^{a_i-1}(p_i - 1)$. Since a finite cyclic group has a unique subgroup of any order dividing the order of the group, it follows that N can be divisible by at most one odd prime. So if N is odd then it must be a power of (an odd) prime. Since $\mathbb{Z}_{2^a}^* \simeq C_2 \times C_{2^{a-2}}$, $a \geq 2$, it follows that N is of the form $1, 2, 4$, or $2 \cdot p^n$ (p an odd prime). Therefore if $\mathbb{Z}_N^*$ is cyclic then N has one of the following values: $1, 2, 4, p^n$, or $2 \cdot p^n$, where p is an odd prime.

4. Mobius μ-function Revisited

In this section, we shall give a characterization of the Mobius μ-function. In order to be able to able to do this we need to discuss some facts about roots of unity and cyclotomic polynomials.

Let n be a positive integer. The polynomial $x^n - 1$ has n zeros. These are $e^{\frac{2\pi i r}{n}} = \cos\frac{2\pi r}{n} + i\sin\frac{2\pi i r}{n}$, $0 \leq r < n$. These n zeros of $x^n - 1$ all lie on the unit circle in the complex plane. As can be easily seen these n zeros of $x^n - 1$ constitute a cyclic group of order n. Those zeros of $x^n - 1$ which have order n (as elements of cyclic group of order n) are called primitive n-th root of unity. The following table gives the primitive roots of unity for $1 \leq n \leq 6$ and $n = 8$.

n	Primitive n-th root of unity
1	1
2	-1
3	$\omega, \omega^2, \qquad \omega = \frac{-1+i\sqrt{3}}{2}$
4	$i, -i$
5	$\alpha, \alpha^2, \alpha^3, \alpha^4, \qquad \alpha = \cos\frac{2\pi}{5} + i\sin\frac{2\pi}{5}$
6	$-\omega, -\omega^2,$
8	$\dfrac{\pm 1 \pm i}{\sqrt{2}}$

The cyclotomic polynomials $\Phi_n(x)$ are defined as follows. Let $\alpha = e^{\frac{2\pi i}{n}} = \cos\frac{2\pi}{n} + i\sin\frac{2\pi}{n}$ be a primitive n-th root of unity. Let $G = \langle\alpha\rangle$, be the cyclic group of order n consisting of the n n-th roots of unity. From group theory we know that $Order(\alpha^r) = \dfrac{n}{(r,n)}$. Thus α^r has order n if and only if $(r,n) = 1$. Thus the $\phi(n)$ elements of G which have order n are precisely the elements α^r, $1 \le r \le n, (r,n) = 1$. The n-th cyclotomic polynomial $\Phi_n(x)$ is defined as

$$\Phi_n(x) = \prod_{\substack{1 \le r \le n \\ (r,n)=1}} (x - \alpha^r)$$

Evidently $\Phi_n(x)$ has degree $\phi(n)$. We shall be using the following well-known properties of the polynomials $\Phi_n(x)$. These can be found in any of the standard text on Abstract Algebra.

(i) $\Phi_n(x)$ has integer coefficients.

(ii) $\Phi_n(x)$ is irreducible over $\mathbb{Z}$.

For every $n \in \mathbb{N}$ we have

$$x^n - 1 = \prod_{d|n} \Phi_d(x)$$

Observe that this identity gives another proof of the fact that $n = \sum_{d|n} \phi(d)$.

An Example

Letting $n = 12$ and $n = 6$ in the above expression for $x^n - 1$ we have (suppressing x in $\Phi_d(x)$)

$$x^{12} - 1 = \Phi_1 \cdot \Phi_2 \cdot \Phi_3 \cdot \Phi_4 \cdot \Phi_6 \cdot \Phi_{12}$$

$$x^6 - 1 = \Phi_1 \cdot \Phi_2 \cdot \Phi_3 \cdot \Phi_4 \cdot \Phi_6$$

Therefore

$x^{12} - 1 = (x^6 - 1)\Phi_4 \cdot \Phi_{12}$ and so $x^6 + 1 = \Phi_4 \cdot \Phi_{12}$. Since $\Phi_4 = x^2 + 1$ we have

$$\Phi_{12} = \frac{x^6 + 1}{x^2 + 1} = x^4 - x^2 + 1$$

If p is a prime number then

$$x^{p^n} - 1 = \prod_{i=0}^{i=n} \Phi_{p^i} = (x^{p^{n-1}} - 1)\Phi_{p^n}$$

Therefore

$$\Phi_{p^n} = \frac{x^{p^n} - 1}{x^{p^{n-1}} - 1} = x^{(p-1)p^{n-1}} + x^{(p-2)p^{n-1}} + \cdots + x^{p^{n-1}} + 1$$

In particular, $\Phi_{27} = x^{18} + x^9 + 1, \quad \Phi_{16} = x^8 + 1, \quad \Phi_{125} = x^{100} + x^{75} + x^{50} + x^{25} + 1$. So the N-the cyclotomic polynomial when N is a power of a prime number is easily determined. The following table gives a list of cyclotomic polynomials for various values of n.

n	$\Phi_n(x)$
1	$x - 1$
2	$x + 1$
3	$x^2 + x + 1$
4	$x^2 + 1$
5	$x^4 + x^3 + x^2 + x + 1$
6	$x^2 - x + 1$
7	$x^6 + x^5 + x^4 + x^3 + x^2 + x + 1$
8	$x^4 + 1$
9	$x^6 + x^3 + 1$
10	$x^4 - x^3 + x^2 - x + 1$
12	$x^4 - x^2 + 1$
14	$x^6 - x^5 + x^4 - x^3 + x^2 - x + 1$
15	$x^8 - x^7 + x^5 - x^4 + x^3 - x + 1$
18	$x^6 - x^3 + 1$
20	$x^8 - x^6 + x^4 - x^2 + 1$
25	$x^{20} + x^{15} + x^{10} + x^5 + 1$

We shall now prove the following theorem which gives a characterization of the Mobius μ-function.

Theorem Let $n \in \mathbb{N}$ and α be a primitive n-th root of unity. Define

$$f(n) = \sum_{\substack{1 \le i \le n \\ gcd(i,n)=1}} \alpha^i$$

which is the sum of primitive n-th roots of unity. Then $f(n) = \mu(n)$.

Proof First we show that f is a multiplicative function. Suppose $gcd(m,n) = 1$. It is clear that if α is a primitive n-th root of unity and β is a primitive m-th root of unity then $\alpha\beta$ is a primitive mn-th root of unity. We have

$$f(n) \cdot f(m) = \sum_{\substack{1 \le i \le n \\ gcd(i,n)=1}} \alpha^i \times \sum_{\substack{1 \le j \le m \\ gcd(j,m)=1}} \beta^j = \sum_{\substack{1 \le i \le n \\ 1 \le j \le m \\ gcd(i,n)=gcd(j,m)=1}} \alpha^i \beta^j = f(nm).$$

The last equality follows from the observation that each summand in the last sum is a primitive mn-th root of unity and there are exactly $\phi(mn)$ distinct summands. This proves that f is a multiplicative function. So we only need to prove the assertion of the theorem when $n = p^k$, a power of the prime number p. We have seen that

$$\Phi_p(x) = x^{p-1} + x^{p-2} + \cdots + x + 1$$

and

$$\Phi_{p^k}(x) = x^{(p-1)p^{k-1}} + x^{(p-2)p^{k-1}} + \cdots + x^{p^{k-1}} + 1$$

Therefore $f(p) = -1$, and $f(p^k) = 0$ if $k > 1$, because the the sum of zeros of a monic polynomial is equal to, except for the sign, the second leading coefficient. Hence $f(p^k) = \mu(p^k)$. Since f is multiplicative function, $f(n) = \mu(n)$ for all n. This completes the proof of the theorem.

CHAPTER 4

Quadratic Residues, Nonresidues, and the Law of Quadratic Reciprocity

Let p be an odd prime, and a be an integer such that $gcd(a,p) = 1$. Then a is said to be a *quadratic residue mod* p if the congruence $x^2 \equiv a(\text{mod } p)$ has a solution, otherwise it is said to be *quadratic nonresidue mod* p. As shown earlier, the group $\mathbb{Z}_p^*$ is cyclic of order $p-1$. The image of the map

$$\begin{array}{ccc} \mathbb{Z}_p^* & \longrightarrow & \mathbb{Z}_p^* \\ x & \longrightarrow & x^2 \pmod p \end{array}$$

is precisely the set of quadratic resides in $\mathbb{Z}_p^*$. Since the kernel of the above map has order two, exactly one half of the elements of $\mathbb{Z}_p^*$ (or any set of $p-1$ nonzero complete residues system mod p) are quadratic residues mod p, and other half are quadratic nonresidues mod p. The quadratic residues mod p can be characterized as the elements of (unique) subgroup of $\mathbb{Z}_p^*$ of order $\frac{p-1}{2}$. For example, if $p = 13$ the set $H = \{1,3,4,9,10,12\}$ consisting of quadratic residues mod p is a subgroup of order 6 of $\mathbb{Z}_{13}^*$. The quadratic nonresidues mod 13 constitute a coset of H in $\mathbb{Z}_{13}^*$. Since for $p > 2$, the square of any number mod p cannot generate $\mathbb{Z}_p^*$, primitive roots mod p are necessarily quadratic nonresidues mod p. But, of course, not every quadratic nonresidue mod p is a primitive root mod p.

1. Legendre Symbol

Let p be an odd prime and a be *any* integer. The *Legendre Symbol* $\left(\frac{a}{p}\right)$ is defined as follows:

$$\left(\frac{a}{p}\right) = \begin{cases} 1 & \text{if a is a quadratic residue mod } p \\ -1 & \text{if a is a quadratic non-residue mod } p \\ 0 & \text{if } p \text{ divides } a \end{cases}$$

Properties of Legendre Symbol

The Legendre symbol satisfies the following properties:

(i) If $a \equiv b(\bmod p)$ then $\left(\frac{a}{p}\right) = \left(\frac{b}{p}\right)$

(ii) $\left(\frac{a}{p}\right) \equiv a^{\frac{p-1}{2}} (\bmod p)$. In particular

$$\left(\frac{-1}{p}\right) = \begin{cases} 1 & \text{if} \quad p \equiv 1 \pmod 4 \\ -1 & \text{if} \quad p \equiv -1 \pmod 4 \end{cases}$$

(iii) $\left(\frac{ab}{p}\right) = \left(\frac{a}{p}\right) \cdot \left(\frac{b}{p}\right)$

Proof (i) Since $a \equiv b \pmod p$, p divides a if and only if p divides b. So if p divides a then $\left(\frac{a}{p}\right) = \left(\frac{b}{p}\right)$. If $p \nmid a$ then $p \nmid b$ and the equality $\left(\frac{a}{p}\right) = \left(\frac{b}{p}\right)$ is an immediate consequence of the definition of quadratic residue and quadratic nonresidue.

(ii) If p divides a then both sides are congruent to zero mod p. Suppose p does not divide a. We write

$$a^{p-1} - 1 = (a^{\frac{p-1}{2}} + 1)(a^{\frac{p-1}{2}} - 1)$$

By Fermat's Little Theorem we have $a^{p-1} - 1 \equiv 0 \pmod p$. Therefore $a^{\frac{p-1}{2}} \equiv \pm 1 \pmod p$. We know that a is a quadratic residue mod p precisely when it belongs to the (unique) subgroup of order $\frac{p-1}{2}$ in $\mathbb{Z}_p^*$. It follows that such an a is a quadratic residue mod p if and only if $a^{\frac{p-1}{2}} \equiv 1 \pmod p$. Therefore

$$a^{\frac{p-1}{2}} \equiv \begin{cases} 1 \pmod p & \text{if a is a quadratic residue mod } p \\ -1 \pmod p & \text{if a is a quadratic non-residue mod } p. \end{cases}$$

Hence

$$\left(\frac{a}{p}\right) \equiv a^{\frac{p-1}{2}} \pmod p$$

This is known as **Euler's Criterion**.

(iii) If $p|ab$ then both sides are equal to zero. Suppose that p does not divide ab. Then using (ii) property, we get

$$\left(\frac{ab}{p}\right) \equiv (ab)^{\frac{p-1}{2}} = a^{\frac{p-1}{2}} \cdot b^{\frac{p-1}{2}} \equiv \left(\frac{a}{p}\right)\left(\frac{b}{p}\right) \pmod p$$

Since $\left(\dfrac{a}{p}\right)$, $\left(\dfrac{b}{p}\right)$ and $\left(\dfrac{ab}{p}\right)$ can only have values ± 1 we have

$$\left(\frac{ab}{p}\right) = \left(\frac{a}{p}\right) \cdot \left(\frac{b}{p}\right)$$

2. Gauss' Lemma

Theorem (Gauss' Lemma)

Let p be an odd prime and $a > 1$ any integer such that $gcd(a, p) = 1$. Consider the $\frac{p-1}{2}$ integers $a, 2a, \cdots, \frac{p-1}{2}a$ and their least nonnegative residues mod p. If n of these residues exceed $\dfrac{p}{2}$ then

$$\left(\frac{a}{p}\right) = (-1)^n$$

Proof Let the residues which exceed $\dfrac{p}{2}$ be denoted by $r_1, r_2, \cdots, r_n$ and the residues which do not exceed $\dfrac{p}{2}$ be denoted by $s_1, s_2, \cdots, s_k$. Since the $\frac{p-1}{2}$ numbers $a, 2a, \cdots, \frac{p-1}{2}a$ are incongruent mod p we have $n + k = \frac{p-1}{2}$.

Consider the $n + k = \dfrac{p-1}{2}$ numbers:

$$s_1, s_2, \cdots, s_k, p - r_1, p - r_2, \cdots, p - r_n$$

These are all in the interval $[1, \frac{p-1}{2}]$ and are distinct modulo p. For if $p - r_i \equiv s_j \pmod{p}$ then $r_i + s_j \equiv 0 \pmod{p}$. Since $r_i \equiv ua \pmod{p}$ and $s_j \equiv va \pmod{p}$ for some integers u and v such that $1 \leq u, v \leq \frac{p-1}{2}$. We have

$$p|(r_i + s_j) \Longrightarrow p|(ua + va) \Longrightarrow p|(u + v)$$

But this is not possible since $1 < u + v < p$. So the $n + k = \dfrac{p-1}{2}$ numbers

$$s_1, s_2, \cdots, s_k, p - r_1, p - r_2, \cdots, p - r_n$$

are nothing but the numbers $1, 2, \cdots, \frac{p-1}{2}$, possibly in some different order. Therefore

$$s_1 \cdot s_2 \cdots s_k \cdot (p - r_1) \cdot (p - r_2) \cdots (p - r_n) = 1 \cdot 2 \cdots \frac{p-1}{2}$$

that is

$$s_1 \cdot s_2 \cdots s_k \cdot r_1 \cdot r_2 \cdots r_n (-1)^n \equiv 1 \cdot 2 \cdots \frac{p-1}{2} \pmod{p}$$

or

$$a, 2a, \cdots, \frac{p-1}{2}a \cdot (-1)^n \equiv 1 \cdot 2 \cdots \frac{p-1}{2} \pmod p$$

this gives $a^{\frac{p-1}{2}} \cdot (-1)^n \equiv 1 \pmod p$ or equivalently $a^{\frac{p-1}{2}} \equiv (-1)^n \pmod p$.
Since by Euler's Criterion we have $\left(\frac{a}{p}\right) \equiv a^{\frac{p-1}{2}} \pmod p$ we have $\left(\frac{a}{p}\right) = (-1)^n$.

Theorem Let p be an odd prime.

(i) If a is an odd integer not divisible by p then $\left(\frac{a}{p}\right) = (-1)^t$, where

$$t = \sum_{j=1}^{\frac{p-1}{2}} \left[\frac{ja}{p}\right]$$

(ii)

$$\left(\frac{2}{p}\right) = (-1)^{\frac{p^2-1}{8}}$$

Proof

(i) Let $r_1, r_2, \cdots, r_n, s_1, s_2, \cdots, s_k$ have the same meaning as in the Gauss's Lemma. From the definition of the greatest integer function it is clear that if m is a positive integer then $\left[\frac{m}{p}\right]$ is simply the quotient when m is divided by p. Therefore, from the definition of $r_i's$ and $s_j's$ and the Euclid's algorithm it follows that for $1 \le j \le \frac{p-1}{2}$ we have

$$j \cdot a = p \cdot \left[\frac{j \cdot a}{p}\right] + \text{ some } r_i \text{ or some } s_j.$$

Therefore, summing from $j = 1$ to $j = \frac{p-1}{2}$ we get

$$\sum_{j=1}^{\frac{p-1}{2}} ja = p \sum_{j=1}^{\frac{p-1}{2}} \left[\frac{ja}{p}\right] + \sum_{j=1}^{n} r_j + \sum_{j=1}^{k} s_j.$$

Since

$$\sum_{j=1}^{\frac{p-1}{2}} ja = a(1 + 2 + \cdots + \frac{p-1}{2}) = a\frac{p^2-1}{8}$$

we have

$$a\frac{p^2-1}{8} = p \sum_{j=1}^{\frac{p-1}{2}} \left[\frac{ja}{p}\right] + \sum_{j=1}^{n} r_j + \sum_{j=1}^{k} s_j. \qquad (1)$$

As noted preciously, the numbers $p-r_1, p-r_2, \cdots, p-r_n, s_1, s_2, \cdots, s_k$ are nothing but the numbers $1, 2, \cdots, \frac{p-1}{2}$ (possibly) in some different order. Since $n+k = \dfrac{p-1}{2}$, we have

$$\sum_{j=1}^{\frac{p-1}{2}} j = \sum_{j=1}^{n}(p-r_j) + \sum_{j=1}^{k} s_j$$

that is

$$\frac{p^2-1}{8} = np - \sum_{j=1}^{n} r_j + \sum_{j=1}^{k} s_j \tag{2}$$

Subtracting (2) from (1) we get

$$\frac{p^2-1}{8}(a-1) = p\left(\sum_{j=1}^{\frac{p-1}{2}}\left[\frac{ja}{p}\right] - n\right) - 2\sum_{j=1}^{n} r_j. \tag{3}$$

The relations (1), (2), and (3) hold for all $a's$ not divisible by p. If a is odd then, considering the above relations modulo 2 we get

$$\sum_{j=1}^{\frac{p-1}{2}}\left[\frac{ja}{p}\right] - n \equiv 0 \pmod 2$$

or

$$t = \sum_{j=1}^{\frac{p-1}{2}}\left[\frac{ja}{p}\right] \equiv n \pmod 2$$

So $(-1)^n = (-1)^t$, and therefore $\left(\dfrac{a}{p}\right) = (-1)^t$. This proves (i). To prove (ii) we substitute $a = 2$ in (3) above. This gives

$$\frac{p^2-1}{8} = p\left(\sum_{j=1}^{\frac{p-1}{2}}\left[\frac{2j}{p}\right] - n\right) - 2\sum_{j=1}^{n} r_j$$

Evidently, $\sum_{j=1}^{\frac{p-1}{2}}\left[\dfrac{2j}{p}\right] = 0$. Therefore we have

$$\frac{p^2-1}{8} = -np - 2\sum_{j=1}^{n} r_j.$$

This gives

$$\frac{p^2-1}{8} \equiv np \equiv n \pmod{2}.$$

Therefore

$$\left(\frac{2}{p}\right) = (-1)^n = (-1)^{\frac{p^2-1}{8}}$$

or equivalently

$$\left(\frac{2}{p}\right) = \begin{cases} 1 & \text{if } p \equiv \pm 1 \pmod{8} \\ -1 & \text{if } p \equiv \pm 3 \pmod{8} \end{cases}$$

3. Law of Quadratic Reciprocity

Theorem (**Law of Quadratic Reciprocity**)

Let p and q be two distinct odd primes. Then

$$\left(\frac{p}{q}\right) \cdot \left(\frac{q}{p}\right) = (-1)^{\frac{p-1}{2} \cdot \frac{q-1}{2}}$$

Proof:

From the previous theorem we know that

$$\left(\frac{q}{p}\right) = (-1)^{t_1} \text{ where } t_1 = \sum_{j=1}^{\frac{p-1}{2}} \left[\frac{jq}{p}\right]$$

$$\left(\frac{p}{q}\right) = (-1)^{t_2} \text{ where } t_2 = \sum_{j=1}^{\frac{q-1}{2}} \left[\frac{jp}{q}\right]$$

Let

$$L_1 = \left\{(x,y) : x, y \in \mathbb{N}, y < \frac{q}{p}x\right\}$$

$$L_2 = \left\{(x,y) : x, y \in \mathbb{N}, y > \frac{q}{p}x\right\}$$

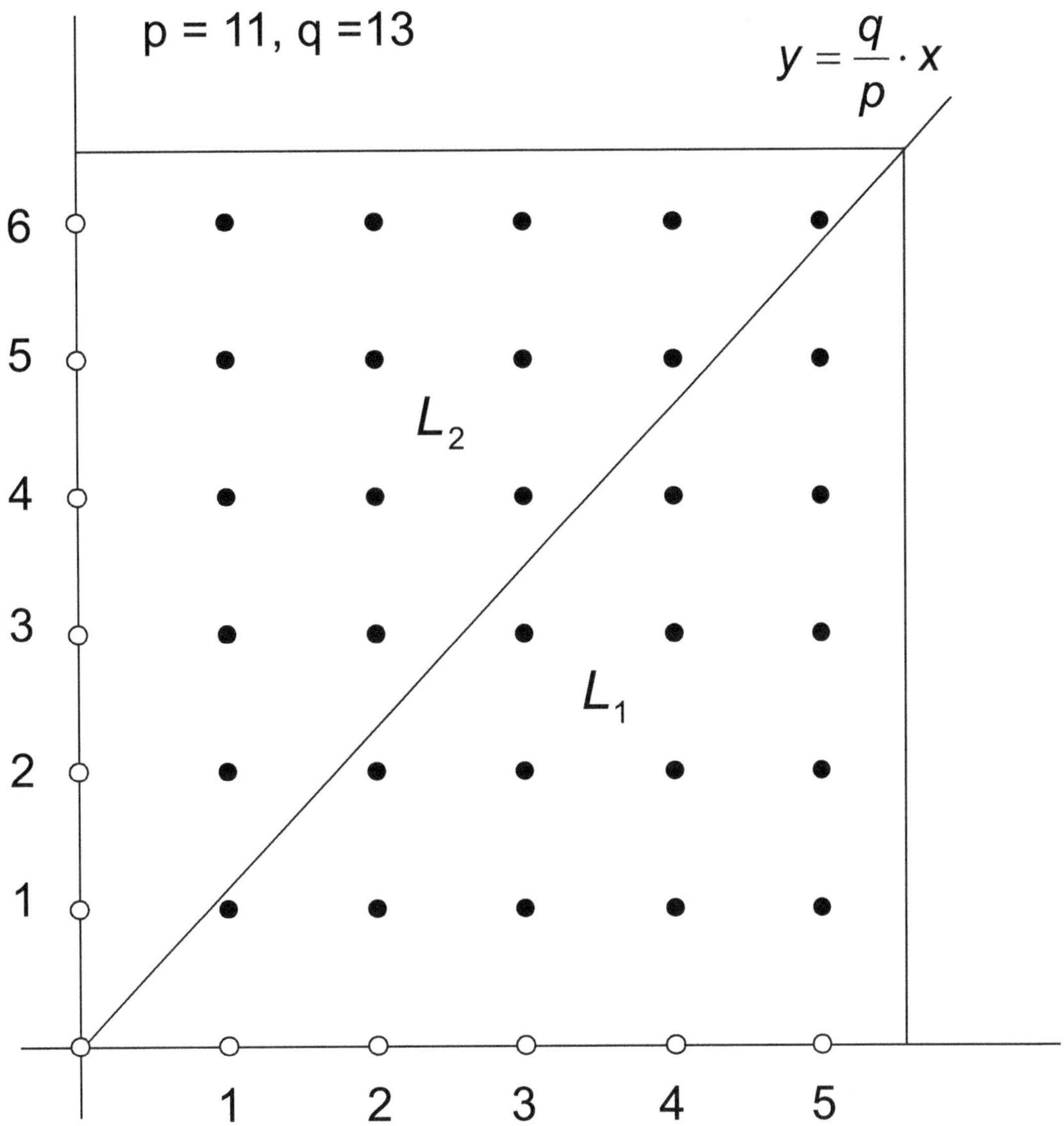

The sets L_1 and L_2 can be regarded as the set of lattice points (points whose both coordinates are integers) in the plane under and over the main diagonal (respectively) $y = \frac{q}{p}x$ of the open rectangle bounded by the lines $x = 0, x = \frac{p}{2}, y = 0, y = \frac{q}{2}$. The adjoining figure shows L_1 and L_2 for the $p = 11$ and $p = 13$. Since there are no lattice points on the diagonal $y = \frac{q}{p}x$, each lattice point in the rectangle lies in L_1 or L_2. For $p = 11$, $q = 13$ we have from the accompanying diagram

$$L_1 = \{(1,1),(2,1),(3,1),(4,1),(5,1),(2,2),(2,3),(2,4),(2,5),(3,3),(3,4),(3,5),(4,4),(4,5),(5,5)\}$$

$$L_2 = \{(1,2),(1,3),(1,4),(1,5),(1,6),(2,3),(2,4),(2,5),(2,6),(3,4),(3,5),(3,6),(4,5),(4,6),(5,6)\}$$

Let L denote the set of all lattice points in the (open) rectangle then $L = L_1 \cup L_2$. Since L_1 and L_2 are disjoint $|L| = |L_1| + |L_2|$. Evidently, $|L| = \frac{p-1}{2} \cdot \frac{q-1}{2}$ We can write

$$L = L_1 \cup L_2 = \left\{(x,y) : x, y \in \mathbb{N}, 1 \leq x \leq \frac{p-1}{2}, 1 \leq y \leq \frac{q-1}{2}\right\}$$

Since

$$L_1 = \left\{(x,y) : x, y \in \mathbb{N}, x, y \geq 1, y < \frac{q}{p}x\right\},$$

we have by counting the lattice points in L_1 lying on the lines parallel to the y-axis

$$|L_1| = \sum_{j=1}^{\frac{p-1}{2}} \#\left\{(j,y) : y \geq 1, y < \frac{q}{p}j\right\} = \sum_{j=1}^{\frac{p-1}{2}} \left[\frac{qj}{p}\right] = t_1$$

Similarly in

$$L_2 = \left\{(x,y) : x, y \in \mathbb{N}, x, y \geq 1, y > \frac{q}{p}x\right\},$$

by counting the lattice points on the lines parallel to x-axis we have

$$|L_2| = \sum_{j=1}^{\frac{q-1}{2}} \#\left\{(x,j) : x \geq 1, x < \frac{p}{q}j\right\} = \sum_{j=1}^{\frac{q-1}{2}} \left[\frac{pj}{q}\right] = t_2$$

Since L is a disjoint union of L_1 and L_2 we have $\frac{p-1}{2} \cdot \frac{q-1}{2} = t_1 + t_2$. Therefore by Gauss' lemma,

$$\left(\frac{p}{q}\right) \cdot \left(\frac{q}{p}\right) = (-1)^{t_1} \cdot (-1)^{t_2} = (-1)^{t_1+t_2} = (-1)^{\frac{p-1}{2} \cdot \frac{q-1}{2}}$$

With the help of the Law of Quadratic Reciprocity (and other properties of the Legendre symbol) we can determine the value of the Legendre symbol $\left(\frac{a}{p}\right)$ for a given value of a and (odd) prime p. An illustration of this follows.

An Example Determine whether the congruence $x^2 \equiv 874 \pmod{1973}$ has a solution. We first note that 1973 is a prime number. Evidently the given problem is equivalent to asking whether 874 is a quadratic residue mod 1973. So we simply have to determine the value of the Legendre symbol $\left(\frac{874}{1973}\right)$. We do as follows:

$$\left(\frac{874}{1973}\right) = \left(\frac{2 \cdot 437}{1973}\right) = \left(\frac{2}{1973}\right) \cdot \left(\frac{437}{1973}\right)$$

Now

$$\left(\frac{2}{1973}\right) = -1 \qquad \text{(since } 1973 \equiv 5 \pmod 8\text{)}$$

$$\left(\frac{437}{1973}\right) = \left(\frac{19 \cdot 23}{1973}\right) = \left(\frac{19}{1973}\right) \cdot \left(\frac{23}{1973}\right)$$

$$\left(\frac{19}{1973}\right) = \left(\frac{1973}{19}\right) = \left(\frac{16}{19}\right) = 1$$

$$\left(\frac{23}{1973}\right) = \left(\frac{1973}{23}\right) = \left(\frac{18}{23}\right) = \left(\frac{2}{23}\right) = 1$$

Therefore, $\left(\frac{874}{1973}\right) = -1$ and the given congruence has no solutions.

4. Jacobi Symbol

Notice that in order to compute the Legendre symbol $\left(\frac{874}{1973}\right)$ we needed to factor 437. We can avoid doing this and thus facilitate the computation of Legendre symbol. In order to do this we shall introduce what is known as *Jacobi symbol.* Let n be a positive odd integer and a be any integer. Let $n = p_1^{r_1} \cdot p_2^{r_2} \cdots p_k^{r_k}$ be the representation of n as a product of prime numbers. Then the Jacobi symbol $\left(\frac{a}{n}\right)_J$ is defined as

$$\left(\frac{a}{n}\right)_J = \left(\frac{a}{p_1}\right)^{r_1} \cdot \left(\frac{a}{p_2}\right)^{r_1} \cdots \left(\frac{a}{p_k}\right)^{r_k}$$

where the symbols $\left(\frac{a}{p_i}\right)$ on the right for $1 \le i \le k$ are Legendre symbols. The subscript J on the symbol on the left symbol indicates that it is a Jacobi symbol.

Remarks

(1) When $n = p$ is an odd prime number then both the Legendre symbol, $\left(\frac{a}{p}\right)$ and the Jacobi symbol $\left(\frac{a}{p}\right)_J$ are defined and are equal. From this point onward we may drop the subscript J from the Jacobi symbol $\left(\frac{a}{p}\right)_J$ since it will be clear from the context which symbol is meant.

(2) If n is not a prime number then $\left(\frac{a}{n}\right)_J = 1$ does not imply that a is a quadratic residue mod n. For example if $n = 21$ then $\left(\frac{5}{21}\right)_J = 1$ but 5 is not a quadratic residue mod 21 as the congruence $x^2 \equiv 5 \pmod{21}$ has no solutions.

The Jacobi symbol has properties similar to those of the Legendre symbol including its own version of the "Law of Quadratic Reciprocity." Below we list the properties of the Jacobi symbol that we shall be using in the sequel.

Properties of Jacobi Symbol

Let m and n be odd positive integers.

(i) If $a \equiv b \pmod{n}$ then $\left(\frac{a}{n}\right)_J = \left(\frac{b}{n}\right)_J$

(ii) $\left(\frac{ab}{n}\right)_J = \left(\frac{a}{n}\right)_J \left(\frac{b}{n}\right)_J$

(iii) $\left(\frac{a}{mn}\right)_J = \left(\frac{a}{m}\right)_J \left(\frac{a}{n}\right)$

(iv) $\left(\frac{-1}{n}\right)_J = (-1)^{\frac{n-1}{2}}$

(v) $\left(\frac{2}{n}\right)_J = (-1)^{\frac{n^2-1}{8}}$

(vi) $\left(\frac{m}{n}\right)_J \left(\frac{n}{m}\right)_J = (-1)^{\frac{m-1}{2} \cdot \frac{n-1}{2}}$

All these properties of the Jacobi symbol are immediate consequence of the corresponding properties of the Legendre symbol and can be easily proved. As mentioned above the computation of the Legendre symbol can be more effectively carried out using the properties of Jacobi symbol. We demonstrate this by computing the Legendre symbol $\left(\frac{874}{1973}\right)$ which occurred in the example in the previous section. We suppress the subscript J when the symbol in consideration is not a Legendre Symbol.

$$\left(\frac{874}{1973}\right) = \left(\frac{2}{1973}\right) \cdot \left(\frac{437}{1973}\right) = -\left(\frac{437}{1973}\right)$$

$$\left(\frac{437}{1973}\right) = \left(\frac{1973}{437}\right) = \left(\frac{225}{437}\right) = \left(\frac{437}{225}\right) = \left(\frac{212}{225}\right)$$

$$\left(\frac{212}{225}\right) = \left(\frac{53}{225}\right) = \left(\frac{225}{53}\right) = \left(\frac{13}{53}\right) = \left(\frac{53}{13}\right) = \left(\frac{1}{13}\right) = 1$$

Therefore $\left(\frac{874}{1973}\right) = -1$, as seen earlier.

It is important to note in the above computation that we have avoided factoring a number (except for factoring out 2 from an even number). This is very helpful since factoring large numbers, from a computational point of view, is considered to be a rather hard and a time consuming problem.

CHAPTER 5

Representation of Integers as Sums of Squares

1. Sums of two squares

In the first section of this chapter, we shall give a criterion for a positive integer to be expressible as a sum of squares of two integers. First we prove the following lemma.

Lemma Let $p > 2$ be a prime number. Suppose that $x^2 + y^2 = kp$, $1 < k < p$. Then there exists integers k_0, x_0, y_0 such that $x_0^2 + y_0^2 = k_0 p$, $1 \leq k_0 < k$.

Proof We have $x^2 + y^2 = kp$, $1 < k < p$. We can write $x = q_1 k + x_1$, $|x_1| \leq \frac{k}{2}$, and $y = q_2 k + y_1$, $|y_1| \leq \frac{k}{2}$. Clearly, $x_1^2 + y_1^2 \leq \frac{k^2}{4} < k^2$ and $x_1^2 + y_1^2 \equiv 0 \pmod{k}$. So we can write $x_1^2 + y_1^2 = k_0 k$, $1 \leq k_0 < k$. Consider the identity

$$(x^2 + y^2)(x_1^2 + y_1^2) = (xx_1 + yy_1)^2 + (xy_1 - x_1 y)^2.$$

Now

$$xx_1 + yy_1 = (q_1 k + x_1)x_1 + (q_2 k + y_1)y_1 = (q_1 x_1 + q_2 y_1)k + x_1^2 + y_1^2 \equiv 0 \pmod{k}$$

$$xy_1 - x_1 y = (q_1 k + x_1)y_1 - x_1(q_2 k + y_1) = (q_1 y_1 - q_2 x_1)k \equiv 0 \pmod{k}.$$

Let $xx_1 + yy_1 = x_0 k$ and $xy_1 - x_1 y = y_0 k$. Substituting in the above identity we get $kp \cdot k_0 k = x_0^2 k^2 + y_0^2 k^2$. This gives $x_0^2 + y_0^2 = k_0 p$, where $1 \leq k_0 < k$. This completes the proof of the lemma

Note The method of proof of the preceding lemma is known as Fermat's Descent Method.

Theorem Let p be an odd prime. Then p can be written as a sum of two squares if and only if $p \equiv 1 \pmod{4}$.

Proof Since $x^2 + y^2 \equiv 0, 1, 2 \pmod{4}$, a prime (in fact any integer) $p \equiv 3 \pmod{4}$ cannot be expressed as a sum of two squares. So we only need to show that a prime $p \equiv 1 \pmod{4}$ can be expressed as a sum of two squares. Let p be a prime, $p \equiv 1 \pmod{4}$. Since -1 is a quadratic residue mod p, we can find x such that $0 < x < \dfrac{p}{2}$ and $1 + x^2 \equiv 0 \pmod{p}$. Since

$1+x^2 < 1+\dfrac{p^2}{4} < p^2$, we can write $1+x^2 = kp$, where $0 < k < p$. Now the preceding lemma implies the conclusion of the theorem.

Now that we know that any prime $p \equiv 1 \pmod 4$ can be written as a sum of two squares we can easily give a characterization of integers which can be written as sum of two squares. The following theorem gives such a characterization.

Theorem A positive integer n can be written as a sum of two squares if and only if each of its prime divisors which is congruent to $3 \pmod 4$ appears in the prime factorization of n with an even exponent.

Proof Assume that the integer n satisfies the hypothesis of the theorem. Since $2 = 1^2+1^2$, from the identity

$$(a^2+b^2)(x^2+y^2) = (ax+by)^2 + (ay-bx)^2$$

and the fact that a prime $p \equiv 1 \pmod 4$ can be written as a sum of two squares it follows that n can be written as a sum of two squares.

Conversely we shall prove that a positive integer n which does not satisfy the hypothesis of the theorem cannot be written as a sum of two squares. Suppose that n is a positive integer such that it has a prime divisor $p \equiv 3 \pmod 4$, which appears in the prime factorization of n with an odd exponent. We shall prove that n cannot be expressed as a sum of two squares. Suppose that $n = x^2 + y^2$. Let $gcd(x, y) = m$. We can write $x = mx_0$, $y = my_0$, where $gcd(x_0, y_0) = 1$ and $n = m^2(x_0^2 + y_0^2)$. Let $n_0 = \dfrac{n}{m^2}$. As the prime p appears in the prime factorization of n_0 with an odd exponent p divides n_0. Now $n_0 = x_0^2 + y_0^2$, p divides n_0, and $gcd(x_0, y_0) = 1$. Since p cannot divide x_0 or y_0, by multiplying the congruence $x_0^2 + y_0^2 \equiv 0 \pmod p$ by the multiplicative inverse of $x_0 \pmod p$, we can write it as $1 + (x_0^{-1}y_0)^2 \equiv 0 \pmod p$ or $1 + z^2 \equiv 0 \pmod p$ where $z = x_0^{-1}y_0$. But then -1 will be a quadratic residue $\pmod p$, which is not possible since $p \equiv 3 \pmod 4$ and $\left(\dfrac{-1}{p}\right) = (-1)^{\frac{p-1}{2}} = -1$. This completes the proof of the theorem.

There is a characterization of integers which can be written as a sum of three squares. Since its proof (in one direction) involves some results from the theory of quadratic forms, we shall not discuss in these notes. We shall only state (without proof) the following theorem giving such a characterization.

Theorem A positive integer n can be written as a sum of three squares if and only if it is *not* of the form $4^a(8k+7)$.

Note In one direction the proof of the theorem is rather obvious. It can be easily seen that an integer of the form $4^a(8k+7)$ cannot be written as a sum of three squares.

2. Sums of four squares

In this section, we shall prove that every positive integer can be expressed as a sum of four squares. This result was conjectured by Euler and proved by Lagrange.

Lagrange's Identity

$$(a_1^2+a_2^2+a_3^2+a_4^2)(b_1^2+b_2^2+b_3^2+b_4^2) =$$

$$(a_1b_1+a_2b_2+a_3b_3+a_4b_4)^2+(a_1b_2-a_2b_1+a_3b_4-a_4b_3)^2+(a_1b_3-a_3b_1+a_4b_2-a_2b_4)^2$$

$$+(a_1b_4-a_4b_1+a_2b_3-a_3b_2)^2$$

In view of the Lagrange's Identity, in order to prove that every positive integer is a sum of four squares it is sufficient to prove that every prime can be written as a sum of four squares. Since we have already shown that a prime $p \equiv 1 \pmod 4$, can be written as a sum of two squares (and therefore as a sum of four squares) we only need to show that every prime $p \equiv 3 \pmod 4$ can be expressed as a sum of four squares, but this assumption does not simplify the proof.

Theorem Every prime (and therefore every positive integer) can be written as a sum of four squares.

We shall first prove a few lemmas.

Lemma 1 Let G be a finite group and H and K be two nonempty subsets of G. Then either $G = H \cdot K$ or $|G| \geq |H| + |K|$.

Proof Suppose that $|G| < |H| + |K|$. Let $g \in G$. We assert that $H^{-1}g \cap K \neq \emptyset$. For otherwise $|G| \geq |H^{-1}g| + |K| = |H| + |K|$. Let $h \in H$ be such that $h^{-1}g \in K$. Then $g \in HK$, and since g is an arbitrary element of G we have $G = HK$.

Lemma 2 Let $\mathbb{F} = \mathbb{F}_q$ be a finite field of q elements. Then every element of F can be written as a sum of two squares in F_q.

Proof In the preceding lemma let $G = \mathbb{F}_q$, an (additive) group of order q. Let $H = K = \{x^2 : x \in \mathbb{F}_q\}$. If q is even then $H = K = \mathbb{F}_q$ and so every element of $\mathbb{F}_q$ is a square. So we may assume that q is odd. Obviously, $|H| = |K| = 1 + \dfrac{q-1}{2} = \dfrac{q+1}{2}$. Therefore

$|H| + |K| = q + 1 > |G|$. By the previous lemma $\mathbb{F}_q = H + K$. Since every element of $H(= K)$ is a square in $\mathbb{F}$, the proof of the lemma is complete.

Lemma 3 Let p be an odd prime. Then there exists integers x, y such that $1+x^2+y^2 \equiv 0 \pmod{p}$.

Proof Let $G = \mathbb{F}_p = \{0, 1, \cdots, p-1\}$ be a finite field of p elements in the preceding lemma. According to the lemma every element of $\mathbb{F}_p$ can be written as a sum of two squares (in $\mathbb{F}_p$). Applying this to the element -1 in $\mathbb{F}_p$ we see that there exist $x, y \in \mathbb{F}_p$ such that $-1 \equiv x^2 + y^2 \pmod{p}$. That is $1 + x^2 + y^2 \equiv 0 \pmod{p}$.

Lemma 4 Let $p > 2$ be a prime number such that there exist integers a_i, $1 \leq i \leq 4$ satisfying $a_1^2 + a_2^2 + a_3^2 + a_4^2 = mp$, $1 < m < p$. Then we can find integers x_i, $1 \leq i \leq 4$ such that $x_1^2 + x_2^2 + x_3^2 + x_4^2 = m_0 p$, $1 \leq m_0 < m$.

Proof We have $a_1^2 + a_2^2 + a_3^2 + a_4^2 = mp$, $1 < m < p$. Suppose that m is even. Then either all the $a_i's$ are even or all of them are odd or two of them odd and other two even. Should two of them be even and two be odd we may assume without loss of generality that a_1, a_2 are even and a_3, a_4 are odd. In any case we can write

$$\left(\frac{a_1 + a_2}{2}\right)^2 + \left(\frac{a_1 - a_2}{2}\right)^2 + \left(\frac{a_3 + a_4}{2}\right)^2 + \left(\frac{a_3 + a_4}{2}\right)^2 = \frac{a_1^2 + a_2^2 + a_3^2 + a_4^2}{2} = \frac{m}{2} \cdot p$$

So applying this process repeatedly if necessary we may assume that $a_1^2+a_2^2+a_3^2+a_4^2 = mp$, $1 \leq m < p$, m odd. If $m = 1$ then we have nothing to prove. So we may assume that $m > 1$. We now have $a_1^2 + a_2^2 + a_3^2 + a_4^2 = mp$, $1 < m < p$, m odd. As before, using the division algorithm we can write $a_i = q_i m + b_i$, $|b_i| < \frac{m}{2}$, $1 \leq i \leq 4$. We claim that not all $b_i's$ are equal to zero. Suppose that $b_i = 0$, $1 \leq i \leq 4$. Then for $1 \leq i \leq 4$ we have

$$m|a_i \Longrightarrow m^2|a_i^2 \Longrightarrow m^2|mp \Longrightarrow m|p$$

which is not possible since $1 < m < p$. So we conclude that at least one of the $b_i's$ is not equal to zero. Since $a_1^2 + a_2^2 + a_3^2 + a_4^2 \equiv b_1^2 + b_2^2 + b_3^2 + b_4^2 \pmod{m}$, therefore $b_1^2 + b_2^2 + b_3^2 + b_4^2 \equiv 0 \pmod{m}$. Since $b_i^2 < \frac{m^2}{4}$, for $1 \leq i \leq 4$, we have $b_1^2 + b_2^2 + b_3^2 + b_4^2 < m^2$. Therefore we can write $b_1^2 + b_2^2 + b_3^2 + b_4^2 = m_0 m$, $1 \leq m_0 < m$.

Using the Lagrange's identity we can write

$$(a_1^2 + a_2^2 + a_3^2 + a_4^2) \cdot (b_1^2 + b_2^2 + b_3^2 + b_4^2) = y_1^2 + y_2^2 + y_3^2 + y_4^2$$

where

$$\begin{aligned} y_1 &= a_1b_1 + a_2b_2 + a_3b_3 + a_4b_4 \\ y_2 &= a_1b_2 - a_2b_1 + a_3b_4 - a_4b_3 \\ y_3 &= a_1b_3 - a_3b_1 + a_4b_2 - a_2b_4 \\ y_4 &= a_1b_4 - a_4b_1 + a_2b_3 - a_3b_2. \end{aligned}$$

Now

$$a_i = q_im + b_i \Longrightarrow \sum_{i=1}^{4} a_ib_i = \sum_{i=1}^{4} q_imb_i + \sum_{i=1}^{4} b_i^2 \equiv 0 \pmod{m}$$

therefore $y_1 \equiv 0 \pmod{m}$. Also

$$a_1b_2 - a_2b_2 = (q_1m + b_1)b_2 - (q_2m + b_2)b_1 = m(q_1b_2 - q_2b_1) \equiv 0 \pmod{m}$$

and

$$a_3b_4 - a_4b_3 = (q_3m + b_3)b_4 - (q_4m + b_4)b_3 = m(q_3b_4 - q_4b_3) \equiv 0 \pmod{m}$$

therefore $y_2 \equiv 0 \pmod{m}$. Similarly $y_3, y_4 \equiv 0 \pmod{m}$.
So for $1 \leq i \leq 4$ we can write $y_i = mx_i$ or $y_i^2 = m^2x_i^2$. Therefore

$$y_1^2 + y_2^2 + y_3^2 + y_4^2 = m^2(x_1^2 + x_2^2 + x_3^2 + x_4^2).$$

Substituting in the Lagrange's identity above we have

$$mp \cdot m_0m = m^2(x_1^2 + x_2^2 + x_3^2 + x_4^2)$$

This gives $x_1^2 + x_2^2 + x_3^2 + x_4^2 = m_0p$, $1 \leq m_0 < m < p$.

Proof of the main Theorem Let $p > 2$ be a prime. By Lemma 3, we can find integers x, y such that $1 + x^2 + y^2 \equiv 0 \pmod{p}$. By replacing x, y by $p - x$, $p - y$ if necessary we may assume that $1 \leq x, y < \frac{p}{2}$. Then $1 + x^2 + y^2 < 1 + \frac{p^2}{4} + \frac{p^2}{4} < p^2$. In other words, we can find integers a_1, a_2, a_3, and a_4 such that $a_1^2 + a_2^2 + a_3^2 + a_4^2 = mp$, $1 \leq m < p$. By Lemma 4, p can be expressed as a sum of four squares. Since $2 = 1^2 + 1^2 + 0^2 + 0^2$, Lagrange's identity implies that every positive integer can be written as a sum of four squares.

CHAPTER 6

Representing primes in the form $x^2 + Ny^2$

We shall now discuss a problem which is similar to (in fact a generalization of) the problem of expressing a positive integer as a sum of two squares. This is the problem of expressing an integer in the form $x^2 + Ny^2$, where N is a square-free positive integer. In view of the identity

$$(x_1^2 + Ny_1^2) \cdot (x_2^2 + Ny_2^2) = (x_1x_2 + Ny_1y_2)^2 + N(x_1y_2 - x_2y_1)^2$$

our main interest would be to investigate what primes can be expressed in the form x^2+Ny^2. For $N = 1$, this problem reduces to a sum of two squares problem, which has been dealt with earlier. We begin with the following important Theorem of Thue. This theorem will be used repeatedly in what follows.

Theorem (Thue) Let p be an odd prime and b be any (fixed) integer not divisible by p. Then we can find integers x and y, $1 \le x, y < \sqrt{p}$, such that $x \equiv \pm by \pmod{p}$ or equivalently $\pm\frac{x}{y} \equiv b \pmod{p}$.

Proof Let

$$\Omega = \{u : u \in \mathbb{Z}, 0 \le u < \sqrt{p}\}$$

Then $|\Omega| = \left[\sqrt{p}\right] + 1$. Consider the mapping

$$\begin{aligned} \Omega \times \Omega &\longrightarrow \{0, 1, \cdots, p-1\} \\ (u, v) &\longrightarrow u + bv \pmod{p} \end{aligned}$$

Since $|\Omega \times \Omega| = ([\sqrt{p}] + 1)^2 > p$, the above mapping is not one-to-one. So there exist $u_1, v_1, u_2, v_2 \in \Omega$ such that $(u_1, v_1) \ne (u_2, v_2)$ and $u_1 + bv_1 \equiv u_2 + bv_2 \pmod{p}$. It can be easily seen that in such a case $u_1 \ne u_2$ and $v_1 \ne v_2$. Suppose that $u_1 > u_2$. We write $u_1 - u_2 \equiv b(v_1 - v_2) \pmod{p}$. Let $x = u_1 - u_2$ and $y = |v_1 - v_2|$. Clearly, $1 \le x, y < \sqrt{p}$. The last congruence now reads

$$x \equiv \pm by \pmod{p}, \text{ or } \pm\frac{x}{y} \equiv b \pmod{p}$$

This completes the proof of the theorem.

Some Applications of Thue's Theorem

We shall now give several examples of the applications of Thue's Theorem. In what follows p will denote an odd prime number—unless stated otherwise.

Example 1 If $p \equiv 1 \pmod 4$ then p can be written as $x^2 + y^2$.

Since $\left(\frac{-1}{p}\right) = 1$, we can find an integer z such that $z^2 \equiv -1 \pmod p$. Since z is not divisible by p, by Thue's theorem, we can find integers x and y, $1 \leq x, y < \sqrt{p}$ such that $z \equiv \pm\frac{x}{y} \pmod p$. That is

$$z^2 \equiv \frac{x^2}{y^2} \pmod p$$

Therefore

$$\frac{x^2}{y^2} \equiv -1 \pmod p$$

or

$$x^2 + y^2 \equiv 0 \pmod p$$

Since $0 < x^2 + y^2 < 2p$, we must have $x^2 + y^2 = p$. This completes the proof of our assertion.

Example 2 If $p \equiv 1, 3 \pmod 8$ then p can be expressed as $x^2 + 2y^2$. If $p \equiv 5, 7 \pmod 8$ then p cannot be so expressed.

First we shall show that $\left(\frac{-2}{p}\right) = 1$ if and only if $p \equiv 1, 3 \pmod 8$. We have

$$\left(\frac{-2}{p}\right) = 1, \quad p \equiv 1 \pmod 4 \Longrightarrow \left(\frac{2}{p}\right) = 1 \Longrightarrow p \equiv 1 \pmod 8$$
$$\left(\frac{-2}{p}\right) = 1, \quad p \equiv 3 \pmod 4 \Longrightarrow \left(\frac{2}{p}\right) = -1 \Longrightarrow p \equiv 3 \pmod 8$$
$$p \equiv 1 \pmod 8 \Longrightarrow \left(\frac{-2}{p}\right) = \left(\frac{2}{p}\right) = 1$$
$$p \equiv 3 \pmod 8 \Longrightarrow \left(\frac{-2}{p}\right) = -\left(\frac{2}{p}\right) = 1.$$

It follows that if $p \equiv 5, 7 \pmod 8$ then p cannot be expressed as $x^2 + 2y^2$. So we only need to prove the first part of the assertion. We can find z such that $z^2 \equiv -2 \pmod p$. By Thue's theorem we can find x and y, satisfying $1 \leq x, y < \sqrt{p}$ and $z \equiv \pm\frac{x}{y} \pmod p$. So as before we have $x^2 + 2y^2 \equiv 0 \pmod p$. Since $1 \leq x, y < \sqrt{p}$, either $x^2 + 2y^2 = p$ or $x^2 + 2y^2 = 2p$. If $x^2 + 2y^2 = 2p$ then x is even. Let $x = 2x_1$. Then $2x_1^2 + y^2 = p$. This proves that a prime $p \equiv 1, 3 \pmod 8$ can be expressed in the form $x^2 + 2y^2$.

Example 3 If $p \equiv 1 \pmod 3$ then p can be written as $x^2 + 3y^2$. If $p \equiv 2 \pmod 3$ then p cannot be written as $x^2 + 3y^2$.

Suppose that $p \equiv 1 \pmod 3$. Then we have

$$p \equiv 1 \pmod 4, \quad p \equiv 1 \pmod 3 \Longrightarrow \left(\frac{-3}{p}\right) = \left(\frac{3}{p}\right) = \left(\frac{p}{3}\right) = 1$$
$$p \equiv 3 \pmod 4, \quad p \equiv 1 \pmod 3 \Longrightarrow \left(\frac{-3}{p}\right) = -\left(\frac{3}{p}\right) = \left(\frac{p}{3}\right) = 1.$$

Therefore $\left(\frac{-3}{p}\right) == 1$, and so we can find z such that $z^2 \equiv -3 \pmod p$. By Thue's theorem we can find integers x, y, $1 \le x, y < \sqrt{p}$, such that $z \equiv \pm\frac{x}{y} \pmod p$. Then $x^2 + 3y^2 \equiv 0 \pmod p$. Since $x^2 + 3y^2 < 4p$, we must have $x^2 + 3y^2 = p$, $2p$, or $3p$. Now $x^2 + 3y^2 = 2p$ is not possible since this would imply that $x^2 \equiv 2 \pmod 3$.

Suppose that $x^2 + 3y^2 = 3p$. Then x is divisible by 3. Let $x = 3x_1$. Then we have $3x_1^2 + y^2 = p$. This completes the proof of the first part. The second part is obvious since $p \equiv 2 \pmod 3$ implies that $\left(\frac{-3}{p}\right) = -1$ which implies that p cannot be expressed as $x^2 + 3y^2$.

Example 4 A prime p can be expressed as $x^2 + 5y^2$ if and only if $p \equiv 1, 9 \pmod{20}$.

Suppose that $x^2 + 5y^2 = p$. Then $\left(\frac{-5}{p}\right) = 1$. Now either $p \equiv 1 \pmod 4$ or $p \equiv 3 \pmod 4$.

$$p \equiv 1 \pmod 4, \text{ and } \left(\frac{-5}{p}\right) = 1 \Longrightarrow \left(\frac{p}{5}\right) = 1 \Longrightarrow p \equiv 1, 4 \pmod 5$$
$$p \equiv 1 \pmod 4 \text{ and } p \equiv 1, 4 \pmod 5 \Longrightarrow p \equiv 1, 9 \pmod{20}.$$
$$p \equiv 3 \pmod 4, \text{ and } \left(\frac{-5}{p}\right) = 1 \Longrightarrow \left(\frac{p}{5}\right) = -1 \Longrightarrow p \equiv 3, 2 \pmod 5$$
$$p \equiv 3 \pmod 4 \text{ and } p \equiv 3, 2 \pmod 5 \Longrightarrow p \equiv 3, 7 \pmod{20}.$$

But

$$x^2 + 5y^2 = p, \text{ and } p \equiv 3, 7 \pmod{20} \Longrightarrow x^2 \equiv 3, 2 \pmod 5 - \text{ which is not possible.}$$

So we have shown that that if $p \not\equiv 1, 9 \pmod{20}$ then it cannot be expressed as $x^2 + 5y^2$. So we assume that $p \equiv 1, 9 \pmod{20}$. Since $\left(\frac{-5}{p}\right) = 1$ we can find z such that $z^2 \equiv -5 \pmod p$. By Thue's theorem we can find x, y, $1 \le x, y < \sqrt{p}$ such that $z \equiv \pm\frac{x}{y} \pmod p$. Therefore $x^2 + 5y^2 \equiv 0 \pmod p$. Since $1 \le x, y < \sqrt{p}$, it follows that $x^2 + 5y^2 = mp$, where $m = 1, 2, 3, 4$ or 5. If $m = 5$ then x is divisible by 5 and by writing $x = 5x_1$ we get $5x_1^2 + y^2 = p$.

If $m = 4$ then $x^2 + 5y^2 = 4p$, hence both x and y are even. By letting $x = 2x_1, y = 2y_1$, we get $x_1^2 + 5y_1^2 = p$. If $m = 2$ or 3 then $x^2 + 5y^2 = mp$ together with the fact that $p \equiv 1, 9 \pmod{20}$ imply that $x^2 \equiv 2, 3 \pmod 5$—which is not possible. This completes the proof.

Example 5 A prime p can be expressed as $x^2 + 6y^2$ if and only if $p \equiv 1, 7 \pmod{24}$. Suppose that $x^2 + 6y^2 = p$. Then $\left(\frac{-6}{p}\right) = 1$. By using the Chinese remainder theorem it can be seen that $\left(\frac{-6}{p}\right) = 1$ if and only if $p \equiv 1, 5, 7$, or $11 \pmod{24}$ as illustrated in the following diagram

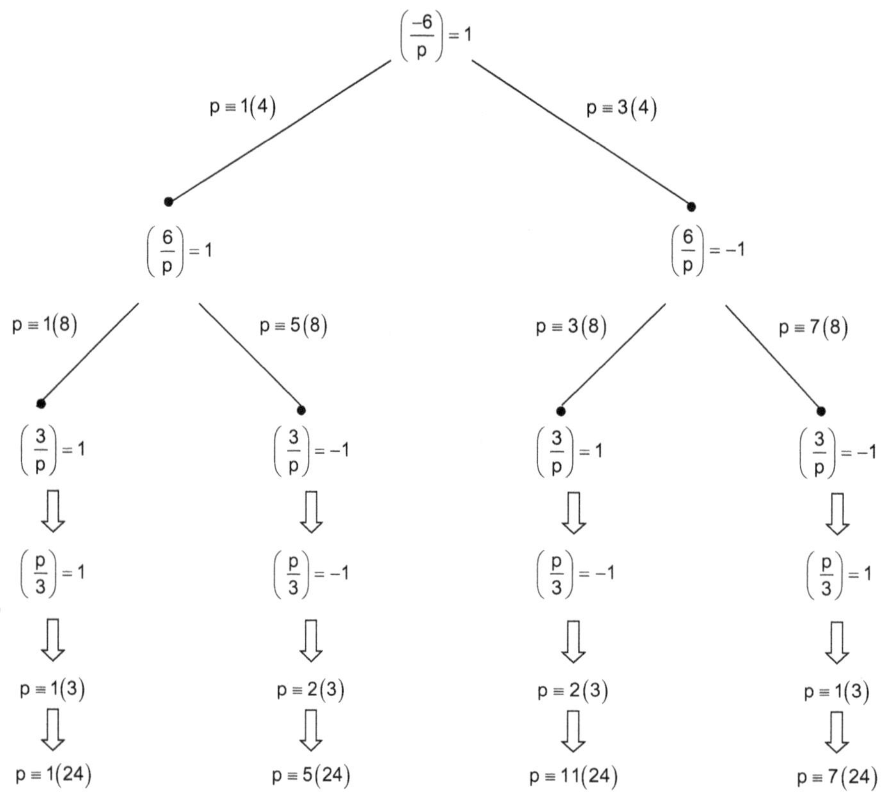

If $x^2 + 6y^2 = p$ and $p \equiv 5, 11 \pmod{24}$ then $x^2 \equiv 2 \pmod{3}$, which is not possible. This shows that if $p \not\equiv 1, 7 \pmod{24}$ then it cannot be expressed as $x^2 + 6y^2$.

We will now show that if $p \equiv 1, 7 \pmod{24}$ then it can be expressed as $x^2 + 6y^2$. Since for such a prime p, $\left(\frac{-6}{p}\right) = 1$, we can find z such that $z^2 \equiv -6 \pmod{p}$. By Thue's theorem we can find x, y, $1 \leq x, y < \sqrt{p}$ such that $z \equiv \pm\frac{x}{y} \pmod{p}$. This give $x^2 + 6y^2 \equiv 0 \pmod{p}$. Since $1 \leq x, y < \sqrt{p}$, we have $x^2 + 6y^2 = mp$, where $m = 1, 2, 3, 4, 5$, or 6. We take these case one at a time.

- If $m = 1$, then $x^2 + 6y^2 = p$ proving our assertion.
- If $m = 2$, then $x^2 + 6y^2 = 2p$ and x must be even. If we let $x = 2x_1$ then $2x_1^2 + 3y^2 = p$. Since $p \equiv 1, 7 \pmod{24}$, it follows that $x_1^2 \equiv 2 \pmod{3}$. This is not possible.
- If $m = 3$, then $x^2 + 6y^2 = 3p$ and x must be divisible by 3. Let $x = 3x_1$. Then $3x_1^2 + 2y^2 = p$. As before, this implies $y^2 \equiv 2 \pmod{3}$, which is not possible.

- If $m = 4$, then $x^2 + 6y^2 = 4p$ and both x and y must be even. Let $x = 2x_1$, $y = 2y_1$. Then $x_1^2 + 6y_1^2 = p$, which proves our assertion.
- If $m = 5$, then $x^2 + 6y^2 = 5p$. Since $p \equiv 1, 7 \pmod{24}$, this gives $x^2 \equiv 2 \pmod 3$. This is not possible.
- If $m = 6$, then $x^2 + 6y^2 = 6p$. In this case x is divisible by 6. Writing $x = 6x_1$ we get $6x_1^2 + y^2 = p$ proving our assertion.

Example 6 If $p \equiv 1, 9, 11, 15, 23, 25 \pmod{28}$ then it can be expressed as $x^2 + 7y^2$. If $p \equiv 3, 5, 13, 17, 19, 27 \pmod{28}$ then it cannot be so expressed.
Suppose that $x^2+7y^2 = p$. Then $\left(\frac{-7}{p}\right) = 1$. By using quadratic reciprocity and the Chinese Remainder Theorem it can be seen that $\left(\frac{-7}{p}\right) = 1$ if and only if $p \equiv 1, 9, 11, 15, 23, 25 \pmod{28}$ as illustrated in the following diagram.
This proves the second part of our assertion. Let $p \equiv 1, 9, 11, 15, 23,$ or $25 \pmod p$. Since $\left(\frac{-7}{p}\right) = 1$, we can find z such that $z^2 \equiv -7 \pmod p$. By Thue's theorem we can find x, y, $1 \le x, y < \sqrt{p}$ such that $z \equiv \pm\frac{x}{y} \pmod p$. This give $x^2 + 6y^2 \equiv 0 \pmod p$. Since $1 \le x, y < \sqrt{p}$, we have $x^2 + 7y^2 = mp$, where $m = 1, 2, 3, 4, 5, 6,$ or 7. We take these case one at a time.

- If $m = 1$, then $x^2 + 7y^2 = p$ proving our assertion.
- If $m = 2$, then $x^2 + 7y^2 = 2p$ which implies that x and y have same parity. This is not possible since $x^2 + 7y^2 \equiv 0 \pmod 4$ and $2p \equiv 2 \pmod 4$.
- If $m = 3$, then $x^2+7y^2 = 3p$. Since neither x nor y can be divisible by 3 considering modulo 3 we have $x^2 + y^2 \equiv 0 \pmod 3$. This is not possible.
- If $m = 4$, then $x^2 + 7y^2 = 4p$. In this case x and y have the same parity. If both x and y are odd then $x^2 + 7y^2$ is a multiple of 8 while $4p$ is not. If both x and y are even then by writing $x = 2x_1$ and $y = 2y_1$, we get $x_1^2 + 7y_1^2 = p$, proving our assertion.
- If $m = 5$, then $x^2 + 7y^2 = 5p$. Then $x^2 + 2y^2 \equiv 0 \pmod 5$, where x and y are not divisible by 5. This is not possible.
- If $m = 6$, then $x^2 + 7y^2 = 6p$. Clearly both x and y must be odd. But then L.H.S. $\equiv 0 \pmod 4$ and R.H.S. $\equiv 2 \pmod 4$ so $m = 6$ is not possible.

This completes the proof.

Another Proof of Fermat's Two Squares Theorem

The following interesting proof of the fact that every prime $p \equiv 1 \pmod 4$ can be expressed as a sum of two squares is due to Don Zagier (*American Mathematical Monthly,* February 1990, page 144).

Let p be a prime, $p \equiv 1 \pmod 4$. Consider the set

$$S = \left\{(x, y, z) \in \mathbb{N}^3 : x^2 + 4yz = p\right\}$$

Evidently S is nonempty. (We can take $x = 1, y = 1$ and $z = \dfrac{p-1}{4}$. We claim that $|S|$ is odd. Consider the following mapping from S to itself.

$$(x, y, z) \longrightarrow \begin{cases} (x + 2z, z, y - x - z) & \text{if} \quad x < y - z \\ (2y - x, y, x - y + z) & \text{if} \quad y - z < x < 2y \\ (x - 2y, x - y + z, y) & \text{if} \quad x > 2y \end{cases}$$

The following can be easily verified.

(i) This mapping is an involution on S (i.e., of order two)

(ii) This mapping has just one fixed point, namely $(1, 1, \frac{p-1}{4})$.

It follows that $|S|$ is odd. We now consider the following permutation (an involution) on S:

$$(x, y, z) \longrightarrow (x, z, y)$$

Since $|S|$ is odd the above mapping has a fixed point. This shows that there exists $(x, y, z) \in N^3$ such that $x^2 + 4yz = p$ and $y = z$. That is, there exist $x, y \in N$ such that $x^2 + 4y^2 = p$. Therefore $p = x^2 + (2y)^2$.

It would be interesting to explore whether such a proof can also be provided for the representation of prime in the form $x^2 + Ny^2$ for some values of $N > 1$.

CHAPTER 7

Infinitude of Primes

Since 2 is the only even prime number, the statement that there are infinitely many primes is equivalently to saying that there are infinitely many odd prime numbers. Consider the polynomial $f(x) = 2x + 1$. Let $p_1, p_2, \cdots, p_k$ be (odd) primes. The (odd) number $f(p_1 \cdot p_2 \cdots p_k) = 2 \cdot p_1 \cdot p_2 \cdots p_k + 1$ is divisible by some (odd) prime which is necessarily different from the primes $p_1, p_2, \cdots, p_k$. This is Euclid's proof of the infinitude of primes which can be stated as saying that there are infinitely many primes of the form $2k + 1$ or equivalently that there are infinitely many primes in the arithmetical progression $1, 3, 5, 7, \cdots, \cdots$. An important theorem of Dirichlet states that any arithmetical progression of natural numbers in which the first term and the common difference are relatively prime contains infinitely many prime numbers. The proof of Dirichlet's theorem for primes in arithmetical progression uses techniques from complex analysis and we do not discuss this proof in this text. We give Euclid-like proofs of the existence of infinitely many primes in various arithmetical progressions and also give a general (algebraic) proof of the fact that there are infinitely many primes of the form $kn + 1$ for a fixed n. In the last section, we give Euler's proof in infinitude of primes.

1. Euclid-like Proofs

Proposition 1: There are infinitely many primes of the form $4k + 1$.

Proof Let $p_1, p_2, \cdots, p_n$ be prime of the form $4k + 1$. Consider the polynomial $f(x) = 4x^2 + 1$. Any prime divisor q of $f(p_1 p_2 \cdots p_n) = 4p_1^2 p_2^2 \cdots p_n^2 + 1$ satisfies $\left(\frac{-1}{q}\right) = 1$. So $q \equiv 1 \pmod 4$ and is obviously different from the primes $p_1, p_2, \cdots, p_n$. So given any finite set of primes of the form $4k + 1$, we can always find another such prime.

Proposition 2: There are infinitely many primes of the form $4k + 3$.

Proof Let $p_1, p_2, \cdots, p_n$ be primes of the form $4k + 3$. Consider the polynomial $f(x) = 4x - 1$. We note that for any integer m, $f(m) \equiv 3 \pmod 4$, so $f(m)$ must have at least one prime divisor of the form $4k+3$. For if every prime divisor of $f(m) \equiv 1 \pmod 4$ then it would imply that $f(m) \equiv 1 \pmod 4$, which is not the case. Let q be a prime, $q \equiv 3 \pmod 4$, which divides $f(p_1p_2\cdots p_n) = 4p_1p_2\cdots p_n - 1$. Obviously, q is different from the primes $p_1, p_2, \cdots, p_n$. So given any finite set of primes of the form $4k+3$, we can find another prime of the form $4k+3$ not in the set.

Proposition 3: There are infinitely many primes of the form $6k+1$.

Proof Let $p_1, p_2, \cdots, p_n$ be primes of the form $6k+1$. Consider the polynomial $f(x) = x^2 - x + 1$. Let $N = f(p_1p_2\cdots p_n) = (p_1p_2\cdots p_n)^2 - (p_1p_2\cdots p_n) + 1$. Then $N \equiv 1 \pmod 6$. Let q be any prime divisor of N. Then $q \neq 3$. Now

$$x^2 - x + 1 \equiv 0 \pmod q \Longrightarrow (2x-1)^2 + 3 \equiv 0 \pmod q \Longrightarrow \left(\frac{-3}{q}\right) = 1 \Longrightarrow q \equiv 1 \pmod 3.$$

As q is odd therefore $q \equiv 1 \pmod 6$. Obviously, q is different from the primes $p_1, p_2, \cdots, p_n$.

Proposition 4: There are infinitely many primes of the form $6k+5$.

Proof Let $p_1, p_2, \cdots, p_n$ be primes of the form $6k+1$. Consider the polynomial $f(x) = 6x - 1$. Let $N = f(p_1p_2\cdots p_n) = 6 \cdot p_1p_2\cdots p_n - 1$. Then $N \equiv 5 \pmod 6$. If every prime divisor p of N were such that $p \equiv 1 \pmod 6$ then we would have $N \equiv 1 \pmod 6$, which is not the case. So N is divisible some prime $q \equiv 5 \pmod 6$ which must be different from the primes $p_1, p_2, \cdots, p_n$.

Proposition 5: There are infinitely many primes of the form $8k+1$.

Proof: Let $f(x) = 16x^4 + 1$. It is enough to show that for any integer x, any prime divisor of $f(x)$ is congruent to $1 \pmod 8$. Let q be prime divisor of $f(x) = 16x^4 + 1$. So

$$16x^4 + 1 \equiv 0 \pmod q \Longrightarrow \left(\frac{-1}{q}\right) = 1 \Longrightarrow q \equiv 1 \pmod 4.$$

Since $16x^4 + 1 = (4x^2+1)^2 - 8x^2$, it follows that $\left(\frac{8}{q}\right) = \left(\frac{2}{q}\right) = 1$. Therefore $q \equiv 1, 7 \pmod 8$. Since $q \equiv 1 \pmod 4$, $q \equiv 7 \pmod 8$ is not possible. So we have $q \equiv 1 \pmod 8$.

Proposition 6: There are infinitely many primes of the form $8k+3$.

Proof Consider the polynomial $f(x) = 4(x^4+x^2)+3$. For any natural number x, $f(x) \equiv 3 \pmod 8$. We show that every prime divisor of $f(x)$ must be congruent to $3 \pmod 8$. Since $f(x) \equiv 3 \pmod 4$, it must have at least one prime divisor which congruent to $3 \pmod 4$.

Let q be any such prime divisor of $f(x)$. The $f(x) = 4(x^4 + x^2) + 3 = (2x^2 + 1)^2 + 2 \equiv 0 \pmod{q}$. So $\left(\frac{-2}{q}\right) = 1$. This together with the fact that $q \equiv 3 \pmod{4}$ implies that $q \equiv 3 \pmod{8}$. Let $p_1, p_2, \cdots, p_n$ be the primes of the form $8k+3$. For $x = p_1 \cdot p_2 \cdots p_n$, $f(x)$ is divisible by some prime $q \equiv 3 \pmod{8}$, and this prime q is different from the primes $p_1, p_2, \cdots, p_n$. This completes the proof.

Caution Note that we also need to show that for $x = p_1 \cdot p_2 \cdots p_n$, $f(x)$ is not a power of 3.

Proof: Let $p_1, p_2, \cdots, p_n$ be the primes of the form $8k+5$. Consider the polynomial $f(x) = x^4 + 2x^2 + 2$. For any odd integer x, $f(x) \equiv 5 \pmod{8}$. We note that $x^4 + 2x^2 + 2 = (x^2+1)^2 + 1$. Let q be any prime divisor of $f(x)$, where $x = p_1 \cdot p_2 \cdots p_n$. Then $\left(\frac{-1}{q}\right) = 1$ and therefore $q \equiv 1 \pmod{4}$. Thus, any prime divisor of $f(x)$ is congruent to 1 or 5 $\pmod{8}$. Since $f(x) \equiv 5 \pmod{8}$ not all of its prime divisors can be congruent to 1 $\pmod{8}$. Hence, $f(x)$ has at least one prime divisor congruent to 5 $\pmod{8}$ and this prime divisor is evidently different from the primes $p_1, p_2, \cdots, p_n$.

Proposition 8: There are infinitely many primes of the form $8k+7$.

Proof: Let $p_1, p_2, \cdots, p_n$ be the primes of the form $8k+7$. Consider the polynomial $f(x) = 8x^2 - 1$. For any integer x, $f(x) \equiv 7 \pmod{8}$. Let q be any prime divisor of $f(x)$, where $x = p_1 \cdot p_2 \cdots p_n$. Then

$$8x^2 - 1 \equiv 0 \pmod{q} \Longrightarrow 16x^2 - 2 \equiv 0 \pmod{q} \Longrightarrow \left(\frac{2}{q}\right) = 1 \Longrightarrow q \equiv 1, 7 \pmod{8}.$$

Since $8x^2 - 1 \equiv 7 \pmod{8}$ not all of its prime divisors can be of the form $8k+1$. So at least one prime divisor of $8x^2 + 1$ is of the form $8k+7$. This prime is evidently different from the primes $p_1, p_2, \cdots, p_n$.

Proposition 9: There are infinitely many primes of the form $12k+1$.

Proof: Consider the polynomial $f(x) = x^4 - x^2 + 1$. We show that for any natural number $x > 1$, any prime divisor q of $f(x)$ is congruent to 1 (mod 12). Let q be a prime divisor of $f(x)$. Then

$$x^4 - x^2 + 1 \equiv 0 \pmod{q} \Longrightarrow 4x^4 - 4x^2 + 4 \equiv 0 \pmod{q} \Longrightarrow (2x^2 - 1)^2 + 3 \equiv 0 \pmod{q}.$$

So $\left(\dfrac{-3}{q}\right) = 1$ which implies that $q \equiv 1$ or $7 \pmod{12}$. Since $(x^2+1)(x^4-x^2+1) = x^6+1$, we have $x^6+1 \equiv 0 \pmod{q}$. So $\left(\dfrac{-1}{q}\right) = 1$ and therefore $q \equiv 1 \pmod{4}$. Hence $q \equiv 1 \pmod{12}$.

Proposition 10: There are infinitely many primes of the form $12k+5$.

Proof: Consider the polynomial $f(x) = 4x^2+1$. Let $p_1, p_2, \cdots, p_n$ be the primes of the form $12k+5$. Let $x = p_1 \cdot p_2 \cdots p_n$. Then $f(x) \equiv 5 \pmod{12}$. Let q be a prime divisor of $f(x)$. Evidently $\left(\dfrac{-1}{q}\right) = 1$. Therefore $q \equiv 1, 5 \pmod{12}$. Since $f(x) \equiv 5 \pmod{12}$, not all prime divisors of $f(x)$ can be congruent to $1 \pmod{12}$. So it has at least one prime divisor congruent to $5 \pmod{12}$ and this prime divisor is evidently different from the primes $p_1, p_2, \cdots, p_n$.

Proposition 11: There are infinitely many primes of the form $12k+7$.

Proof: Consider the polynomial $f(x) = 4x^2+3$. Let $p_1, p_2, \cdots, p_n$ be the primes of the form $12k+7$. Let $x = p_1 \cdot p_2 \cdots p_n$. Then $f(x) \equiv 7 \pmod{12}$. Let q be a prime divisor of $f(x)$. Evidently $\left(\dfrac{-3}{q}\right) = 1$. Therefore $q \equiv 1$ or $7 \pmod{12}$. Since $f(x) \equiv 7 \pmod{12}$, not all prime divisors of $f(x)$ can be congruent to $1 \pmod{12}$. So it has at least one prime divisor congruent to $7 \pmod{12}$ and this prime divisor is evidently different from the primes $p_1, p_2, \cdots, p_n$.

Proposition 12: There are infinitely many primes of the form $12k+11$.

Proof: Consider the polynomial $f(x) = 12x^2-1$. Let $p_1, p_2, \cdots, p_n$ be primes of the form $12k+11$. Let $x = p_1 \cdot p_2 \cdots p_n$. Then $f(x) \equiv 11 \pmod{12}$. Let q be a prime divisor of $f(x)$. Evidently $q \neq 3$. Now

$$12x^2 - 1 \equiv 0 \pmod{q} \Longrightarrow 36x^2 - 3 \equiv 0 \pmod{q} \Longrightarrow \left(\frac{3}{q}\right) = 1.$$

So $q \equiv 1, 11 \pmod{12}$. Since $f(x) \equiv 11 \pmod{12}$, not all prime divisors of $f(x)$ can be $\equiv 1 \pmod{12}$. Therefore, $f(x)$ has at least one prime divisor of the form $12k+11$. This prime divisor is evidently different from the primes $p_1, p_2, \cdots, p_n$.

The proofs of the infinitude of primes in certain arithmetical progressions outlined above are all very similar in essence to the proofs of the infinitude of primes given by Euclid. For this reason we have called them *Euclid-like proofs*. The following table gives the list of polynomials used for the proofs of the infinitude of primes in various arithmetical progressions. It is known that such proofs of infinitude of primes in the arithmetical progression $kn + d$ where $gcd(n, d) = 1$ can be given if and only if $d^2 \equiv 1 \pmod n$. See the book, *The Book of Prime Number Records* by Ribenboim (1972) and references given there.

Euclid-like Proofs

Arithmetical Progression	Polynomial
$2k + 1$	$2x + 1$
$4k + 1$	$4x^2 + 1$
$4k + 3$	$4x - 1$
$6k + 1$	$x^2 - x + 1$
$6k + 5$	$6x - 1$
$8k + 1$	$16x^4 + 1$
$8k + 3$	$4(x^4 + x^2) + 3$
$8k + 5$	$x^4 + 2x^2 + 2$
$8k + 7$	$8x^2 - 1$
$12k + 1$	$x^4 - x^2 + 1$
$12k + 5$	$4x^2 + 1$
$12k + 7$	$4x^2 + 3$
$12k + 11$	$12x^2 - 1$

2. Primes $p \equiv 1 \pmod n$

We now prove the following special case of Dirichlet's theorem on primes in arithmetical progression.

Theorem: Let $n > 1$ be a natural number. Then there are infinitely many primes $p \equiv 1 \pmod n$.

Proof: We know that

$$x^n - 1 = \prod_{d|n} \Phi_d(x)$$

where $\Phi_d(x)$ denotes the dth cyclotomic polynomial. It is known that $\Phi_d(x) \in \mathbb{Z}[x]$ and it is a monic polynomial. For a later use, we note that the constant term of each cyclotomic polynomial $\Phi_d(x)$, for $d > 1$ is equal to 1. This is so because for $d = 2$, $\Phi_2(x) = x + 1$, and for $d > 2$, $\Phi_d(x)$ has even degree and if α is a root of $\Phi_d(x)$ then so is α^{-1} and $\alpha \neq \alpha^{-1}$. Suppose that $p_1, p_2, \cdots, p_r$ are primes congruent to 1 (mod n). Since $\Phi_n(x)$ is a monic polynomial, we can find an integer $t \geq 1$ such that $\Phi_n(n \cdot t \cdot p_1 \cdot p_2 \cdots p_r) > 1$. Let $m = n \cdot t \cdot p_1 \cdot p_2 \cdots p_r$. Let q be a prime divisor of $\Phi_n(m)$. Since q does not divide m, q is not in the set $\{p_1, p_2, \cdots, p_r\}$. Since $\Phi_n(x)$ divides $x^n - 1$, we have $m^n - 1 \equiv 0 \pmod{q}$ and so $Ord_q(m)$, the order of m mod q divides, n. Since $n \cdot m^{n-1} \not\equiv 0 \pmod{q}$, m is not a multiple zero (mod q) of the polynomial $x^n - 1$. So, m is not a zero (mod q) of any of the factors on the right-hand side of

$$x^n - 1 = \prod_{d|n} \Phi_d(x)$$

other than $\Phi_n(x)$. This means that if $d < n$, $d|n$, then m is not a zero (mod q) of $x^d - 1$. Therefore $Ord_q(m) = n$. But $Ord_q(m)$ evidently divides $q - 1$. Therefore n divides $q - 1$ and hence $q \equiv 1 \pmod{n}$. This proves that given any finite set of primes (including an empty set) each congruent to 1 (mod n), we can find another such prime. Hence, there are infinitely many primes congruent to 1 (mod n)

Note: It may be argued that we have assumed the existence of (some) primes congruent to 1 (mod n). But by taking $m = nt$ (for some large t), it can be shown that any prime divisor of $\Phi_n(m)$ is congruent to 1 (mod n). The argument for this is precisely the same as outlined above.

3. Euler's Proof of Infinitude of Primes

In this section, we give Euler's proof of the infinitude of primes.

We will need some elementary facts from the theory of infinite series and infinite products. These can be found in any book on infinite series such as Bromwich (1991) or Knopp (1961). We state below (without proofs) some results from infinite series and infinite products which we use later in this section.

1. If $\sum_{n=0}^{\infty} a_n$ and $\sum_{n=0}^{\infty} b_n$ are two convergent series of positive terms then the series $\sum_{n=0}^{\infty} c_n$, where

$$c_n = \sum_{\substack{i+j=n \\ 0 \leq i, j \leq n}} a_i b_j$$

converges, and

$$\sum_{n=0}^{\infty} a_n \cdot \sum_{n=0}^{\infty} b_n = \sum_{n=0}^{\infty} c_n.$$

2. Terms of an absolutely convergent series can be rearranged in any manner without affecting the sum of the series.
3. The series $\sum_{n=1}^{\infty} \frac{1}{n^s}$ for $s > 1$ represents a continuous function of s and its sum is denoted by $\zeta(s)$.

Lemma: $\lim_{s \to 1+} \zeta(s) = \infty$.

Proof: Consider the graph, of the function $y = \frac{1}{x^s}$, $s > 1$.

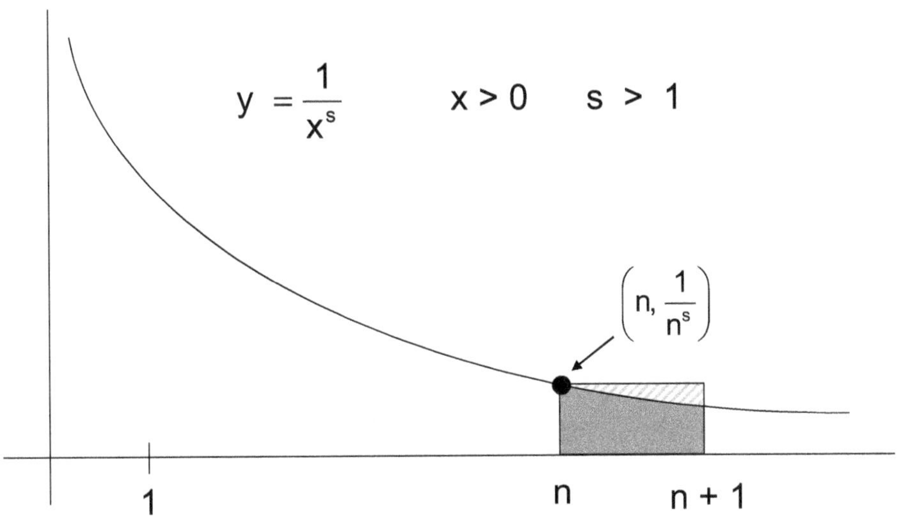

Area under the curve $y = \frac{1}{x^s}$ from $x = n$ to $x = n + 1$ is $\int_n^{n+1} \frac{1}{x^s} dx$

Area of the rectangular part is $\frac{1}{n^s}$, $\frac{1}{n^s} > \int_n^{n+1} \frac{1}{x^s} dx$

From this graph it is evident that

$$\frac{1}{n^s} > \int_n^{n+1} \frac{1}{x^s} \, dx.$$

Therefore

$$\sum_{n=1}^{\infty} \frac{1}{n^s} > \int_1^{\infty} \frac{1}{x^s} \, dx = \frac{1}{s-1}, \qquad \text{for } s > 1.$$

That is

$$\zeta(s) > \frac{1}{s-1}, \qquad \text{for} \quad s > 1.$$

Therefore

$$\lim_{s\to 1+} \zeta(s) \geq \lim_{s\to 1+} \frac{1}{s-1} = \infty.$$

Theorem: There are infinitely many prime numbers.

Proof: (Euler)

Suppose that there are only finitely many prime numbers $p_1, p_2, \cdots, p_k$. For any positive s we write

$$\frac{1}{1-\left(\dfrac{1}{p_i}\right)^s} = \sum_{m=0}^{\infty} \frac{1}{p_i^{ms}}.$$

We shall be needing it only for $s > 1$. Since every natural number can be uniquely expressed as a product of prime numbers and since (by our assumption) $p_1, p_2, \cdots, p_k$ are all the distinct prime numbers, it follows that for $s > 1$:

$$\prod_{i=1}^{k} \frac{1}{1-p_i^{-s}} = \prod_{i=1}^{k} \left(\sum_{m=0}^{\infty} \frac{1}{p_i^{ms}} \right) = \sum_{n=1}^{\infty} \frac{1}{n^s}.$$

In the last equality, we have used the fundamental theorem of arithmetic and the results about multiplying series of positive terms and on rearrangement of the terms of an absolutely convergent series. It follows that for $s > 1$:

$$\zeta(s) = \prod_{i=1}^{k} (1 - p_i^{-s})^{-1}.$$

$$\lim_{s\to 1+} \zeta(s) = \prod_{i=1}^{k} (1 - p_i^{-1})^{-1}.$$

Since the right-hand side is finite, it contradicts the statement of the previous lemma. Hence, we conclude that there are infinitely many primes.

CHAPTER 8

Pell's Equation and Continued Fractions

Diophantine equations of the type $x^2 - dy^2 = 1$, where $d > 1$ is a square-free positive integer, are known as Pell's equation in number theory. It is now well known that Euler had mistakenly attributed such equations to Pell although the English mathematician Pell had nothing to do with them. We shall show that such equations have infinitely many solutions and once we have found the "least" of these solutions the others can easily be obtained. The following sequence of lemmas will be used to show the existence of a solution of a Pell's equation.

Lemma 1

If θ is an irrational number then the set

$$\{a + b\theta : a, b \in \mathbb{Z}\}$$

is a dense subset of the set $\mathbb{R}$ of all real numbers.

Proof:

It is enough to show that, given a natural number n, there exist integers a and b, not both zero, such that $|a + b\theta| < \frac{1}{n}$. Consider the $n+1$ numbers $z_i = i\theta - [i\theta]$, which are fractional parts of the numbers $i\theta$, for $1 \leq i \leq n+1$. By the Pigeon Hole Principle there exist integers i, j, $1 \leq i < j \leq n+1$, such that $|z_i - z_j| < \frac{1}{n}$. That is,

$$|i\theta - [i\theta] - (j\theta - [j\theta])| < \frac{1}{n}$$

or

$$|[j\theta] - [i\theta] + (i - j)\theta| < \frac{1}{n},$$

That is, $|a + b\theta| < \frac{1}{n}$, where $a = [j\theta] - [i\theta]$, and $b = i - j$.

Remark: (i) If θ is positive then for inequality to be valid a and b must have opposite signs. Also, since $i \neq j$, we have $|b| = j - i \leq n$.

(ii) Let $\theta = \sqrt{d}$, where $d > 1$ be a square-free positive integer, then the above lemma implies that, given n however large, there exist positive integers x and y such that, $y \leq n$ and $|x - y\sqrt{d}| < \frac{1}{n}$. Since $y \leq n$, we have $\frac{1}{n} \leq \frac{1}{y}$. So $|x - y\sqrt{d}| < \frac{1}{y}$. We shall use this in the proof of the next lemma.

Lemma 2

Let $d > 1$ be a square-free integer. Then there are infinitely many $x, y \in \mathbb{N}$ satisfying

$$|x - y\sqrt{d}| < \frac{1}{y}$$

Proof:

We have already seen that there is at least one such pair of natural numbers. It would suffice to show that given one such pair (x, y) we can find another pair (x_1, y_1) of natural numbers satisfying

(i) $|x_1 - y_1\sqrt{d}| < |x - y\sqrt{d}|$

(ii) $|x_1 - y_1\sqrt{d}| < \frac{1}{y_1}$

Let n_1 be a natural number such that $n_1 > \frac{1}{|x - y\sqrt{d}|}$. In view of the remarks following the preceding lemma we can find a pair (x_1, y_1) of natural numbers such that $y_1 \leq n_1$ and $|x_1 - y_1\sqrt{d}| < \frac{1}{n_1}$.

The pair (x_1, y_1) has the required properties, since

(i) $|x_1 - y_1\sqrt{d}| < \frac{1}{n_1} < |x - y\sqrt{d}|$

(ii) $|x_1 - y_1\sqrt{d}| < \frac{1}{n_1} \leq \frac{1}{y_1}$

Lemma 3 Let $d > 1$ be a square-free integer. Then there exist infinitely many $x, y \in \mathbb{N}$ satisfying

$$|x^2 - dy^2| < 2\sqrt{d} + 1$$

Proof: By previous lemma, there exist infinitely many $x, y \in \mathbb{N}$ such that $|x - y\sqrt{d}| < \frac{1}{y}$. For such x and y we have

$$|x + y\sqrt{d}| = |x - y\sqrt{d} + 2y\sqrt{d}| \leq |x - y\sqrt{d}| + 2y\sqrt{d} < \frac{1}{y} + 2y\sqrt{d},$$

so

$$|x^2 - dy^2| = |x + y\sqrt{d}| \cdot |x - y\sqrt{d}| < \left(\frac{1}{y} + 2y\sqrt{d}\right) \cdot \frac{1}{y} = \frac{1}{y^2} + 2\sqrt{d} \leq 1 + 2\sqrt{d}$$

Corollary There exists a natural number m such that there are infinitely many $x, y \in \mathbb{N}$ such that $|x^2 - dy^2| < m$.

Proof: It follows immediately by letting $m = 2 + [2\sqrt{d}]$ in the statement of the previous lemma.

From this corollary we get

Lemma 4 Let $d > 1$ be a square free positive integer. There exists an integer k such that the Diophantine equation $x^2 - dy^2 = k$ has infinitely many solutions.

We shall now prove the main theorem of this section.

Theorem:

If $d > 1$ is a square-free integer then there exists integers x and $y(\neq 0)$ such that $x^2 - dy^2 = 1$.

Proof

By Lemma 4, there exists an integer k such that the Diophantine equation $x^2 - dy^2 = k$ has infinitely many solutions. Consider the following set

$$S = \left\{(x, y), x, y \in \mathbb{N} : x^2 - dy^2 = k, \right\}$$

We define a relation $\sim$ on S as follows. If $(x_1, y_1), (x_2, y_2) \in S$ then $(x_1, y_1) \sim (x_2, y_2)$ if $x_1 \equiv x_2 \pmod{|k|}$ and $y_1 \equiv y_2 \pmod{|k|}$. As can be easily checked, this defines an equivalence relation on S. Under this equivalence relation, there are only finitely many (in fact at most $|k|$) equivalence classes. So there is at least one equivalence class with infinitely many elements. In particular, we can find two distinct pairs of positive integers (x_1, y_1) and (x_2, y_2) such that $x_1 \equiv x_2 \pmod{|k|}$, $y_1 \equiv y_2 \pmod{|k|}$, satisfying $x_1 - dy_1^2 = x_2 - dy_2^2 = k$. We write

$$\frac{x_1 + y_1\sqrt{d}}{x_2 + y_2\sqrt{d}} = A + B\sqrt{d}$$

where A and B are rational numbers. We shall show that A and B are non-zero integers and $A^2 - B^2 d = 1$. We write

$$\begin{aligned} A + B\sqrt{d} = \frac{x_1 + y_1\sqrt{d}}{x_2 + y_2\sqrt{d}} &= \frac{x_1 + y_1\sqrt{d}}{x_2 + y_2\sqrt{d}} \cdot \frac{x_2 - y_2\sqrt{d}}{x_2 - y_2\sqrt{d}} \\ &= \frac{(x_1x_2 - y_1y_2d) + (x_2y_1 - x_1y_2)\sqrt{d}}{x_2^2 - y_2^2\sqrt{d}} \\ &= \frac{x_1x_2 - y_1y_2d}{k} + \frac{x_2y_1 - x_1y_2}{k}\sqrt{d} \end{aligned}$$

Now

$$x_1x_2 - y_1y_2d \equiv x_1^2 - y_1^2 d \equiv 0 \pmod{|k|}$$

and

$$x_2y_1 - x_1y_2 \equiv x_2y_2 - x_2y_2 = 0 \pmod{|k|}$$

Therefore,

$$A = \frac{x_1x_2 - y_1y_2d}{k}, \qquad B = \frac{x_2y_1 - x_1y_2}{k}$$

are integers. As

$$\left.\begin{array}{rcl} A + B\sqrt{d} & = & \dfrac{x_1 + y_1\sqrt{d}}{x_2 + y_2\sqrt{d}} \\ A - B\sqrt{d} & = & \dfrac{x_1 - y_1\sqrt{d}}{x_2 - y_2\sqrt{d}} \end{array}\right\} \Longrightarrow A^2 - B^2d = \frac{x_1^2 - y_1^2d}{x_2^2 - y_2^2d} = 1$$

we have $A^2 - B^2d = 1$. In order that the pair (A, B) constitutes a nontrivial solution of the Pell's equation $x^2 - dy^2 = 1$, we must show that $B \neq 0$. Suppose that $B = 0$. Then $x_2y_1 - x_1y_2 = 0$. Therefore, $\dfrac{x_2}{x_1} = \dfrac{y_2}{y_1} = t(\text{say})$. So $x_2 = x_1t$ and $y_2 = y_1t$. We have in $x_2^2 - y_2^2d = t^2(x_1^2 - y_1^2d) = k$. This gives $t^2 = 1$ and therefore $t = \pm 1$. Since both $x_i's$ and $y_i's$ are positive integers we conclude that $t = 1$. This gives $x_2 = x_1$ and $y_2 = y_1$. This contradicts the fact that the pairs (x_1, y_1) and (x_2, y_2) are distinct. Hence $B \neq 0$. This completes the proof of the theorem.

So far, we have only established that the Diophantine equation $x^2 - dy^2 = 1$ has at least one solution. In fact, this equation has infinitely many solutions and given the "least solution" (soon to be defined) we can describe them all. If $(\alpha, \beta) \in \mathbb{Z} \times \mathbb{Z}$ is a solution of $x^2 - dy^2 = 1$ then so are $(\pm\alpha, \pm\beta)$. To each solution (α, β) of $x^2 - dy^2 = 1$ satisfying $\alpha > 0, \beta > 0$, we can associate the real number $\alpha + \beta\sqrt{d}$. We shall now prove the following theorem.

Theorem

There exists a solution (a, b) of the Diophantine equation $x^2 - dy^2 = 1$, such that any solution (α, β) satisfying $\alpha > 0, \beta > 0$ is given by

$$\alpha + \beta\sqrt{d} = (a + b\sqrt{d})^n$$

where n is some natural number.

Proof

Let (a, b) denote a solution to the Diophantine equation $x^2 - dy^2 = 1$, such that $a > 0, b > 0$ and a is the least possible. Such a solution exists and is unique. Let $(\alpha, \beta) \in \mathbb{N} \times \mathbb{N}$ be another solution of $x^2 - dy^2 = 1$ in positive integers. Clearly $a + b\sqrt{d} > 1$ and $\alpha + \beta\sqrt{d} > a + b\sqrt{d}$.

If $\alpha+\beta\sqrt{d}$ is not an integral power of $a+b\sqrt{d}$ then there is a unique positive integer n such that

$$(a+b\sqrt{d})^n < \alpha+\beta\sqrt{d} < (a+b\sqrt{d})^{n+1},$$

or

$$1 < \frac{\alpha+\beta\sqrt{d}}{(a+b\sqrt{d})^n} < a+b\sqrt{d}$$

we can write

$$1 < \frac{\alpha+\beta\sqrt{d}}{(a+b\sqrt{d})^n} = (\alpha+\beta\sqrt{d})(a-b\sqrt{d})^n = A+B\sqrt{d}$$

where A and B are some integers. We note that $A+B\sqrt{d} < a+b\sqrt{d}$ and $A^2-B^2d=1$. Since $A+B\sqrt{d}>1$ and from the four numbers $\pm A \pm B\sqrt{d}$, the largest one is the only one that exceeds 1 and has both rational and irrational parts positive, we must have $A>0$ and $B>0$. Now since $A+B\sqrt{d} < a+b\sqrt{d}$ and $A^2-B^2d = a^2-b^2d = 1$ and A, B, a, b are all positive we must have $A<a$ (and $B<b$). This contradicts the choice of $a+b\sqrt{d}$. Hence $\alpha+\beta\sqrt{d}$ must be positive integral power of $a+b\sqrt{d}$.

Now that we have shown that the Pell's equation $x^2-dy^2=1$ does have a nontrivial solution (in fact infinitely many of them) we proceed to discuss a method of finding the "least solution" (and therefore all of them).

If d is small then all we have to do is to find the least positive integer y such that $1+dy^2$ is a perfect square. Sometimes (but not always) it is easier to find a least y such that $-1+dy^2$ is a perfect square. In such a case the "least solution" of the Pell's equation is (x_1, y_1) where $x_1+y_1\sqrt{d} = (x+y\sqrt{d})^2$, where $x^2-y^2d=-1$ and y is the least positive integer such that $-1+dy^2$ is a perfect square. For example, for $d=13$, we have $18^2-13\cdot 5^2=-1$. So the "least solution" (x_1, y_1) of $x^2-13y^2=1$ is given by $x_1+y_1\sqrt{13}=(18+5\sqrt{13})^2$, which gives $x_1=649$ and $y_1=180$.

This method is feasible only for relatively small values of d. For $d=94$ the "least solution" of the Pell's equation $x^2-94y^2=1$ is given by $x=2,143,295$, $y=221,064$. The usual method of finding the least solution of the Pell's equation $x^2-dy^2=1$ is by means of continued fractions. In the next section we shall discuss the properties of continued fractions. Our main purpose for doing so is to give an algorithm to determine the "least" solution of the Pell's equation.

1. Continued Fractions

Consider the process of the determination of the GCD of the numbers 149 and 63 using the Euclidean algorithm. We write

$$149 = 2 \cdot 63 + 23 \qquad \text{so} \qquad \frac{149}{63} = 2 + \frac{23}{63} = 2 + \frac{1}{63/23}$$

$$63 = 2 \cdot 23 + 17 \qquad \text{so} \qquad \frac{63}{23} = 2 + \frac{17}{23} = 2 + \frac{1}{23/17}$$

$$23 = 1 \cdot 17 + 6 \qquad \text{so} \qquad \frac{23}{17} = 1 + \frac{6}{17} = 1 + \frac{1}{17/6}$$

$$17 = 2 \cdot 6 + 5 \qquad \text{so} \qquad \frac{17}{6} = 2 + \frac{5}{6} = 2 + \frac{1}{6/5}$$

$$6 = 1 \cdot 5 + 1 \qquad \text{so} \qquad \frac{6}{5} = 1 + \frac{1}{5}$$

Therefore we can write

$$\frac{149}{63} = 2 + \cfrac{1}{2 + \cfrac{1}{1 + \cfrac{1}{2 + \cfrac{1}{1 + \cfrac{1}{5}}}}}$$

or more compactly in the commonly used notation as

$$\frac{149}{63} = 2 + \frac{1}{2+}\,\frac{1}{1+}\,\frac{1}{2+}\,\frac{1}{1+}\,\frac{1}{5}$$

The expression on the right is known as the *continued fraction expansion* of the number on the left. It is clear from above that if we start with a (positive) rational number its continued fraction always terminates. Continued fraction expansion of an irrational number can be determined in a similar fashion. It can be seen that the continued fraction expansion of an irrational number does not terminate. For example, let $\alpha = \sqrt{14}$. We write

$$\begin{aligned}
\sqrt{14} &= 3 + (\boldsymbol{\sqrt{14} - 3}) \\
\frac{1}{\sqrt{14}-3} &= \frac{\sqrt{14}+3}{5} &&= 1 + \frac{(\sqrt{14}-2)}{5} \\
\frac{5}{\sqrt{14}-2} &= \frac{5(\sqrt{14}+2)}{10} &&= \frac{\sqrt{14}+2}{2} &&= 2 + \frac{\sqrt{14}-2}{2} \\
\frac{2}{\sqrt{14}-2} &= \frac{\sqrt{14}+2}{5} &&= 1 + \frac{\sqrt{14}-3}{5} \\
\frac{5}{\sqrt{14}-3} &= \sqrt{14}+3 &&= 6 + (\boldsymbol{\sqrt{14} - 3})
\end{aligned}$$

So we can write

$$\sqrt{14} = 3 + \cfrac{1}{1 + \cfrac{1}{2 + \cfrac{1}{1 + \cfrac{1}{6+}}}} \cdots$$

or

$$\sqrt{14} = 3 + \frac{1}{1+}\;\frac{1}{2+}\;\frac{1}{1+}\;\frac{1}{6+}\cdots$$

or simply as

$$\sqrt{14} = (3; \overline{1, 2, 1, 6})$$

where the bar indicates that the terms under it repeat indefinitely.

1.1. Partial Quotients and Convergents.

Consider the continued fraction

$$a_0 + \frac{1}{a_1+}\;\frac{1}{a_2+}\;\frac{1}{a_3+}\cdots$$

where $a_i's$ are natural numbers. These $a_i's$ are called *partial quotients.* If we terminate a continued fraction at a specific partial quotient the resulting fractions are called *convergents.* So the first three convergents for the above continued fraction are:

$$\frac{a_0}{1}, \qquad \frac{a_0a_1+1}{a_1}, \qquad \frac{a_0(a_1a_2+1)+a_2}{a_1a_2+1}$$

The numerators and denominators of the convergents are generally denoted as $p_i's$ and $q_i's$. In the case of the continued fraction for $\sqrt{14}$, the initial convergents are:

$$\frac{p_0}{q_0} = \frac{3}{1}, \qquad \text{so} \quad p_0 = 3, \quad q_0 = 1$$

$$\frac{p_1}{q_1} = 3 + \frac{1}{1} = \frac{4}{1}, \qquad \text{so} \quad p_1 = 4, \quad q_1 = 1$$

$$\frac{p_2}{q_2} = 3 + \cfrac{1}{1 + \cfrac{1}{2}} = \frac{11}{3}, \qquad \text{so} \quad p_2 = 11, \quad q_2 = 3$$

$$\frac{p_3}{q_3} = 3 + \cfrac{1}{1 + \cfrac{1}{2 + \cfrac{1}{1}}} = \frac{15}{4}, \qquad \text{so} \quad p_3 = 15, \quad q_3 = 4$$

$$\frac{p_4}{q_4} = 3 + \cfrac{1}{1 + \cfrac{1}{2 + \cfrac{1}{1 + \cfrac{1}{6}}}} = \frac{101}{27}, \qquad \text{so} \quad p_4 = 101, \quad q_4 = 27$$

So the first five convergents for the continued fraction of $\sqrt{14}$ are:

$$\frac{3}{1}, \ \frac{4}{1}, \ \frac{11}{3}, \ \frac{15}{4}, \ \frac{101}{27}$$

2. Complete Quotients

Let

$$\alpha = a_0 + \frac{1}{a_1+} \ \frac{1}{a_2+} \ \frac{1}{a_3+} \cdots$$

be the continued fraction of a number α. The complete quotients $\alpha's$ of the continued fraction of α are the numbers which are represented by the continued fraction by starting at some partial quotient. So

$$\alpha_0 = \alpha = \ a_0 + \frac{1}{a_1+} \ \frac{1}{a_2+} \ \frac{1}{a_3+} \cdots$$

$$\alpha_1 = a_1 + \frac{1}{a_2+}\,\frac{1}{a_3+}\cdots$$

$$\alpha_2 = a_2 + \frac{1}{a_3+}\cdots$$

So, for $\alpha = \dfrac{149}{63}$ the complete quotients are:

$$\alpha_0 = \frac{149}{63}, \quad \alpha_1 = \frac{63}{23}, \quad \alpha_2 = \frac{23}{17}, \quad \alpha_3 = \frac{17}{6}, \quad \alpha_4 = \frac{6}{5}, \quad \text{and } \alpha_5 = 5,$$

as can be seen from the continued fraction expansion of $\dfrac{149}{63}$ given earlier. It is clear that if α is irrational then so is each of the complete quotients in its continued fraction. For example, in case of $\alpha = \sqrt{14}$, whose continued fraction is:

$$\alpha = \sqrt{14} = (3; \overline{1, 2, 1, 6}) = 3 + \frac{1}{1+}\,\frac{1}{2+}\,\frac{1}{1+}\,\frac{1}{6+}\cdots$$

the complete quotients are:

$$\alpha_0 = \alpha = \sqrt{14}, \quad \alpha_1 = \frac{\sqrt{14}+3}{5}, \quad \alpha_2 = \frac{\sqrt{14}+2}{2}, \quad \alpha_3 = \frac{\sqrt{14}+2}{5}, \quad \alpha_4 = \sqrt{14}+3.$$

After this the last four complete quotients for the continued fraction of $\sqrt{14}$ repeat indefinitely. For the continued fraction:

$$a_0 + \frac{1}{a_1+}\,\frac{1}{a_2+}\,\frac{1}{a_3+}\cdots$$

we define p_n and q_n $(n \geq 0)$ recursively as follows:

$$p_0 = a_0, \quad q_0 = 1, \quad p_1 = a_0a_1 + 1, \quad q_1 = a_1, \quad \text{and for } n \geq 2,$$

$$p_n = a_np_{n-1} + p_{n-2}, \quad q_n = a_nq_{n-1} + q_{n-2}.$$

Theorem For $n \geq 0$, $\dfrac{p_n}{q_n}$, where p_n and q_n are as defined above, constitute the n-th convergents of the above continued fraction.

Proof: We shall prove the theorem by induction. The result evidently holds for $n = 0$. Assume that the result is true for some $n \geq 0$. So the n-th convergent for the above continued fraction is

$$\frac{p_n}{q_n} = \frac{a_np_{n-1} + p_{n-2}}{a_nq_{n-1} + q_{n-2}}$$

It is clear from the definition and the process of forming convergents that the $(n+1)$-th convergent is obtained from the n-th convergents by replacing a_n by $a_n + \dfrac{1}{a_{n+1}}$. Therefore, the $(n+1)$-th convergent for the above continued fraction is given by

$$\begin{aligned}
\frac{\left(a_n + \dfrac{1}{a_{n+1}}\right) p_{n-1} + p_{n-2}}{\left(a_n + \dfrac{1}{a_{n+1}}\right) q_{n-1} + q_{n-2}} &= \frac{a_n p_{n-1} + p_{n-2} + \dfrac{p_{n-1}}{a_{n+1}}}{a_n q_{n-1} + q_{n-2} + \dfrac{q_{n-1}}{a_{n+1}}} \\
&= \frac{a_{n+1}(a_n p_{n-1} + p_{n-2}) + p_{n-1}}{a_{n+1}(a_n q_{n-1} + q_{n-2}) + q_{n-1}} \\
&= \frac{a_{n+1} p_n + p_{n-1}}{a_{n+1} q_n + q_{n-1}} \\
&= \frac{p_{n+1}}{q_{n+1}}
\end{aligned}$$

Hence the $(n+1)$-th convergent is of the desired form.

The next theorem will help us to see that p_n and q_n are relatively prime.

Theorem: Let $\dfrac{p_n}{q_n}$ be the n-th convergent for the continued fraction

$$a_0 + \frac{1}{a_1+}\,\frac{1}{a_2+}\,\frac{1}{a_3+}\cdots$$

Then

$$p_n q_{n-1} - p_{n-1} q_n = (-1)^{n-1}, \qquad \text{for} \quad n \geq 1.$$

In particular, $gcd(p_n, q_n) = 1$.

Proof: We shall prove the theorem by induction. For $n = 1$, we have

$$p_1 q_0 - p_0 q_1 = (a_0 a_1 + 1) \cdot 1 - a_0 a_1 = 1 = (-1)^0.$$

So the theorem is true for $n = 1$. Assume that the theorem is true for some $n \geq 0$. We have

$$\begin{aligned}
p_{n+1} q_n - p_n q_{n+1} &= (a_{n+1} p_n + p_{n-1}) q_n - p_n (a_{n+1} q_n + q_{n-1}) \\
&= -(p_n q_{n-1} - p_{n-1} q_n) = (-1) \cdot (-1)^{n-1} = (-1)^n.
\end{aligned}$$

As our main purpose in studying continued fractions is to give a method of determining "least" solution of the Pell's equation $x^2 - dy^2 = 1$, we shall need to study the continued fraction of some special kind of quadratic irrationals, which, following Davenport (1993) we shall call *reduced quadratic irrational.*

3. Reduced Quadratic Irrational

Definition An algebraic number α is said to be a *reduced quadratic irrational* if

(i) α is real, $\alpha > 1$ and as an algebraic number is of degree two.

(ii) Its algebraic conjugate $\alpha^{'}$ satisfies $-1 < \alpha^{'} < 0$.

For example, $1+\sqrt{2}$ and $1+\sqrt{3}$ are reduced quadratic irrationals whereas $1+\sqrt{5}$ is not. If α is a reduced quadratic irrational then obviously $\alpha+\alpha^{'} > 0, \quad \alpha-\alpha^{'} > 0$, and $\alpha\alpha^{'} < 0$. So if $\alpha = a+b\sqrt{d}$, where $a, b \in \mathbb{Q}$ is a reduced quadratic irrational then $a > 0$ and $b > 0$. Therefore, for a reduced quadratic irrational α if $ax^2+bx+c \in \mathbb{Z}[x]$ is such that $a\alpha^2+b\alpha+c = 0$, then $\dfrac{b}{a} < 0$ and $\dfrac{c}{a} < 0$. So if we take a to be positive then both b and c are negative.

Shape of a Reduced Quadratic Irrational and Periodicity of its Continued Fraction Expansion

Let α be an algebraic number. Let $f(x) \in \mathbb{Z}[x]$ denote the unique polynomial satisfying the following conditions.

(i) $f(\alpha) = 0$.

(ii) $f(x)$ is irreducible over $\mathbb{Z}$ - in particular the GCD of its coefficients is 1.

(iii) The leading coefficient of $f(x)$ is positive

It is easy to see that an algebraic number α uniquely determines such a polynomial. We shall denote by $I(\alpha)$ the unique polynomial determined by the algebraic number α, satisfying the above conditions. Let α be a reduced quadratic irrational and $\alpha^{'}$ be its algebraic conjugate. Suppose that the corresponding polynomial $I(\alpha)$ is ax^2+bx+c. Then

$$\alpha = \frac{-b+\sqrt{b^2-4ac}}{2a}, \qquad \alpha^{'} = \frac{-b-\sqrt{b^2-4ac}}{2a}$$

Note that we have taken the positive sign in front of the radical in the expression for α because we know that the rational and irrational parts of a reduced quadratic irrational are both positive. This also implies that $a > 0$, $b < 0$, $c < 0$ and since α is real algebraic number of degree two, $b^2 - 4ac > 0$ is not a perfect square. Let $P = -b$, $Q = 2a$, and $D = b^2 - 4ac$. We can write

$$\alpha = \frac{P + \sqrt{D}}{Q}, \qquad \alpha^{'} = \frac{P - \sqrt{D}}{Q}$$

Since $\alpha^{'} < 0$ we have $P < \sqrt{D}$. Also as $\alpha > 1$ we have $P + \sqrt{D} > Q$. That is $Q < P + \sqrt{D} < 2\sqrt{D}$. In what follows we shall be making a frequent use of these inequalities. Note that $\dfrac{P + \sqrt{D}}{Q}$ may not be in reduced form - but we shall ignore the possible cancelation and continue to write α as $\dfrac{P + \sqrt{D}}{Q}$. For example, if α is a reduced quadratic irrational such that $I(\alpha) = 7x^2 - 8x - 3$, then

$$\alpha = \frac{8 + \sqrt{148}}{14} = \frac{4 + \sqrt{37}}{7}$$

but still in this case we shall let $P = 8$, $Q = 14$ and $D = 148$.

Let us try to expand a reduced quadratic irrational α into continued fractions in the usual way. We write

$$\alpha = [\alpha] + \alpha - [\alpha] = a_0 + \frac{1}{\alpha_1}$$

where $a_0 = [\alpha]$ and $\alpha_1 = \dfrac{1}{\alpha - [\alpha]} = \dfrac{1}{\alpha - a_0}$. So $\alpha^{'} = a_0 + \dfrac{1}{\alpha_1^{'}}$ and

$$\alpha_1^{'} = \frac{1}{\alpha^{'} - a_0} = -\frac{1}{a_0 - \alpha^{'}}.$$

Since $-1 < \alpha^{'} < 0$, we get $a_0 = [\alpha] > 1$, $a_0 - \alpha^{'} > 1$. Therefore, $-1 < \alpha_1^{'} < 0$. As $\alpha_1 > 1$, it follows that α_1 is also a reduced quadratic irrational. This shows that as we try to expand a reduced quadratic irrational α into continued fraction, then not only the first complete quotient but each subsequent complete quotient is also a reduced quadratic irrational. Since $\alpha_1 = \dfrac{1}{\alpha - a_0}$, it follows that the the polynomials $I(\alpha)$ and $I(\alpha_1)$ have the same discriminant. That is if $I(\alpha) = ax^2 + bx + c$ and $I(\alpha_1) = a_1x^2 + b_1x + c_1$, then $b^2 - 4ac = b_1^2 - 4a_1c_1$. So we can write $\alpha_1 = \dfrac{P_1 + \sqrt{D}}{Q_1}$, where as observed before P_1 and Q_1 are positive integers satisfying $P_1 < \sqrt{D}$ and $Q_1 < 2\sqrt{D}$. Since this is true for each complete quotient it follows that the pair (P_i, Q_i) has only finitely many possibilities. This in turn shows that there are only finitely many possible complete quotients. Therefore, the continued fraction expansion of a reduced quadratic irrational is periodic.

An Example

The following table illustrates this for the reduced quadratic irrational $\alpha = 7 + \sqrt{62}$.

α	α_1	$I(\alpha)$	P	Q	$I(\alpha_1)$
$7+\sqrt{62}$	$\frac{7+\sqrt{62}}{13}$	$x^2-14x-13$	14	2	$13x^2-14x-1$
$\frac{7+\sqrt{62}}{13}$	$\frac{6+\sqrt{62}}{2}$	$13x^2-14x-1$	14	26	$2x^2-12x-13$
$\frac{6+\sqrt{62}}{2}$	$\frac{6+\sqrt{62}}{13}$	$2x^2-12x-13$	12	4	$13x^2-12x-2$
$\frac{6+\sqrt{62}}{13}$	$7+\sqrt{62}$	$13x^2-12x-2$	12	26	$x^2-14x-13$

We now proceed to show that the continued fraction expansion of a reduced quadratic irrational is *purely periodic*, in another words it is periodic from the very beginning.

Theorem: Let α be a reduced quadratic irrational. Then the continued fraction expansion of α is purely periodic, that is, its period begins at the first partial quotient.

Proof: Let α_i denote the i-th complete quotient of α. To prove the theorem it is enough to show that in the continued fraction expansion of α, if two complete quotients are equal then so are the two preceding complete quotients. Let $m, n \in \mathbb{N}$, $m \neq n$ and α_m and α_n be the m-th and n-th complete quotients in the continued fraction expansion of α and $\alpha_m = \alpha_n$. We shall show that $\alpha_{m-1} = \alpha_{n-1}$. For any index $i > 0$ we have

$$\alpha_i = [\alpha_i] + \frac{1}{\alpha_{i+1}} = a_i + \frac{1}{\alpha_{i+1}} \qquad \text{and therefore} \qquad \alpha_i' = a_i + \frac{1}{\alpha_{i+1}'}$$

Let

$$\beta_i = -\frac{1}{\alpha_i'}, \qquad \beta_{i+1} = -\frac{1}{\alpha_{i+1}'}$$

Therefore,

$$\alpha_i' = -\frac{1}{\beta_i}, \qquad \alpha_{i+1}' = -\frac{1}{\beta_{i+1}}.$$

We note that as α is a reduced quadratic irrational so are all the complete quotients $\alpha_i's$. Since $-1 < \alpha_i' < 0$, $\beta_i = -\frac{1}{\alpha_i'} > 1$. Also as $\beta_i' = -\frac{1}{\alpha_i} > -1$, we have $-1 < \beta_i' < 0$. So $\beta_i's$ are also reduced quadratic irrationals.
Substituting in $\alpha_i' = a_i + \frac{1}{\alpha_{i+1}'}$ we have

$$-\frac{1}{\beta_i} = a_i - \beta_{i+1}, \qquad \text{or} \qquad \beta_{i+1} = a_i + \frac{1}{\beta_i}$$

So $[\beta_{i+1}] = a_i$ for $i \geq 0$. Substituting $i = m-1$ and $i = n-1$ we have

$$\beta_m = a_{m-1} + \frac{1}{\beta_{m-1}} = [\beta_m] + \frac{1}{\beta_{m-1}}, \qquad \beta_n = a_{n-1} + \frac{1}{\beta_{n-1}} = [\beta_n] + \frac{1}{\beta_{n-1}}$$

Clearly

$$\alpha_m = \alpha_n \Longrightarrow \beta_m = \beta_n \Longrightarrow a_{m-1} = a_{n-1}.$$

Now as

$$\alpha_{m-1} = a_{m-1} + \frac{1}{\alpha_m}, \qquad \alpha_{n-1} = a_{n-1} + \frac{1}{\alpha_n}$$

$\alpha_m = \alpha_n$ and $a_{m-1} = a_{n-1} \Longrightarrow \alpha_{m-1} = \alpha_{n-1}$. This completes the proof of the theorem.
We conclude that the period of the continued fraction of a reduced quadratic irrational α begins at $a_0 = [\alpha]$. Therefore, the continued fraction expansion of a reduced quadratic irrational α is purely periodic.

Since our main purpose in studying the continued fraction is to provide an algorithm for solving Pell's equation $x^2 - dy^2 = 1$, we shall now study the continued fraction expansion of $\sqrt{N}$ for $N \in \mathbb{N}$, where $N > 1$ is not a perfect square. We shall need the continued fraction expansion of $\sqrt{N}$ for the algorithm. We shall first prove a lemma concerning certain specific types of reduced quadratic irrationals.

Lemma Let N be a natural number which is not a perfect square. Let $\alpha = \sqrt{N} + a_0$, where $a_0 \in \mathbb{N}$. Then α is a reduced quadratic irrational if and only if $a_0 = [\sqrt{N}]$.

Proof Evidently, if $a_0 = [\sqrt{N}]$ then α is a reduced quadratic irrational. Suppose that $\alpha = \sqrt{N} + a_0$ is a reduced quadratic irrational. Then by the definition of a reduced quadratic irrational $-1 < \alpha^{'} < 0$, where $\alpha^{'}$ is the algebraic conjugate of α. In our case $\alpha^{'} = -\sqrt{N} + a_0$. Therefore,

$$-1 < -\sqrt{N} + a_0 < 0,$$

so

$$a_0 < \sqrt{N} < a_0 + 1$$

Therefore, $a_0 = [\sqrt{N}]$. This completes the proof of the lemma.

Let $N > 1$ be a natural number which is not a perfect square. Let $\alpha = \sqrt{N} + a_0$, where $a_0 = [\sqrt{N}]$. So $[\alpha] = 2a_0$. By previous lemma, α is a reduced quadratic irrational. We write

$$\alpha = [\alpha] + (\alpha - [\alpha]) = 2a_0 + \frac{1}{\alpha_1}, \qquad \alpha_1 = \frac{1}{\alpha - [\alpha]}$$

We know that the continued fraction expansion of α is purely periodic. So we can write

$$\begin{aligned} \alpha &= 2a_0 + \frac{1}{a_1+}\ \frac{1}{a_2+}\ \frac{1}{a_3+} \cdots \frac{1}{a_n+}\ \frac{1}{2a_0+} + \cdots \\ &= (\overline{2a_0; a_1, a_2, a_3, \cdots, a_n, 2a_0}) \end{aligned}$$

Therefore,

$$\begin{aligned} \sqrt{N} = \alpha - a_0 &= a_0 + \frac{1}{a_1+}\ \frac{1}{a_2+}\ \frac{1}{a_3+} \cdots \frac{1}{a_n+}\ \frac{1}{2a_0+} + \cdots \\ &= (a_0, \overline{a_1, a_2, a_3, \cdots, a_n, 2a_0}) \end{aligned}$$

The bars indicate the periodic part of the continued fraction. We shall show that $a_i \le a_0$ for $1 \le i \le n$. This means that while expanding $\sqrt{N}$ as a continued fraction we could stop when we get $2a_0 = 2[\sqrt{N}]$ as one of the partial quotients, because after that subsequent partial quotients are mere repetition of the previous partial quotients. Since $\alpha = \sqrt{N} + a_0$, where $a_0 = [\sqrt{N}]$, the polynomial $I(\alpha)$ is given by

$$I(\alpha) = x^2 - 2a_0 x + (a_0^2 - N).$$

We write

$$\alpha = \frac{2a_0 + \sqrt{4N}}{2} = \frac{P + \sqrt{D}}{Q}$$

where $P = 2a_0$, $Q = 2$, $D = 4N$. As noted before, the first complete quotient $\alpha_1 = \dfrac{1}{\alpha - a_0}$. Hence the discriminant of the polynomial $I(\alpha)$ equals the discriminant of the polynomial $I(\alpha_1)$. Since the middle coefficient of $I(\alpha)$ is even, so is the middle coefficient of $I(\alpha_1)$, and therefore, of each $I(\alpha_i)$ for $i \geq 1$. Thus all the $P_i's$ for $i \geq 1$ are even. Evidently each Q_i is even, as it is equal to twice the leading coefficients of the $I(\alpha_i)$. Since $Q_i's$ are positive we have $Q_i \geq 2$. At the i-th step in the development of continued fraction we have

$$\alpha_i = a_i + \frac{1}{\alpha_{i+1}},$$

and

$$\alpha_i = \frac{P_i + \sqrt{D}}{Q_i}$$

Now

$$a_i < \alpha_i = \frac{P_i + \sqrt{D}}{Q_i} < \frac{2\sqrt{D}}{Q_i}$$

If $Q_i = 2$ then $\alpha_i = \dfrac{P_i + \sqrt{D}}{2} = \dfrac{P_i}{2} + \sqrt{N}$ (since $D = 4N$), and as α_i is a reduced quadratic irrational, we have $\dfrac{P_i}{2} = [\sqrt{N}]$, that is, $P_i = 2a_0$. So, in this case, $\alpha_i = \alpha$ and we have $i = 0$ or $i = n+1$. It follows that for $1 \leq i \leq n$, $Q_i > 2$, as $Q_i's$ are all even we have $Q_i \geq 4$. Therefore for $1 \leq i \leq n$ we have

$$a_i < \alpha_i = \frac{P_i + \sqrt{D}}{Q_i} < \frac{2\sqrt{D}}{Q_i} \leq \frac{4\sqrt{N}}{4} = \sqrt{N}$$

That is, $a_i \leq [\sqrt{N}]$ for $1 \leq i \leq n$ or $a_i \leq a_0$. This shows that if $N > 1$ is not a perfect square then $\sqrt{N}$ has a continued fraction of the form,

$$\sqrt{N} = (a_0; \overline{a_1, a_2, \cdots, a_n, 2a_0})$$

where $a_0 = [\sqrt{N}]$ and $a_i \leq a_0$ for $1 \leq i \leq n$.

In an appendix we have given a list of continued fraction expansion of $\sqrt{N}$, where N is not a perfect square, for $2 \le N \le 500$. Below we summarize the results that we have proved concerning the continued fraction expansion of a quadratic irrational $\sqrt{N}$, where $N > 1$ is an integer which is not a perfect square.

(i) The continued fraction of $\sqrt{N}$ (where N is not a perfect square) has the form

$$\sqrt{N} = (a_0; \overline{a_1, a_2, \cdots, a_n, 2a_0})$$

where $a_i \le a_0$ for $1 \le i \le n$. This continued fraction has period $n+1$.

(ii) The numbers $a_0, a_1, a_2, \cdots$ are known as partial quotients.

(iii) The complete quotients $\alpha_k' s$ are given by

$$\alpha_k = a_k + \frac{1}{a_{k+1}+} \frac{1}{a_{k+2}+} \cdots$$

(iv) The truncated continued fraction

$$a_0 + \frac{1}{a_1+} \frac{1}{a_2+} \frac{1}{a_3+} \cdots \frac{1}{a_n}$$

is called the n-th convergent and is generally written as the rational number $\frac{p_n}{q_n}$.

(v) The numerators and denominators of the n-th convergent $\frac{p_n}{q_n}$ can be (recursively) computed as follows:

$$p_0 = a_0, q_0 = 1, \quad p_1 = a_0 a_1 + 1, \quad q_1 = a_1, \quad \text{and} \quad \text{for} \quad n \ge 2$$

$$p_n = a_n p_{n-1} + p_{n-2}$$

$$q_n = a_n q_{n-1} + q_{n-2}.$$

For $n \ge 1$, these satisfy the relation

$$p_n q_{n-1} - p_{n-1} q_n = (-1)^{n-1}$$

We shall now describe a method of solving Pell's equation $x^2 - Ny^2 = 1$ where $N \in \mathbb{N}$ is a not a perfect square.

4. Solution of the Pell's Equation $x^2 - Ny^2 = 1$

Let $N > 1$ be a natural number which is not a perfect square. Suppose that the continued fractions expansion of $\sqrt{N}$ has period $n+1$. Consider the $(n+1)$-th convergent of the continued fraction expansion of $\sqrt{N}$. It is given by

$$\frac{p_{n+1}}{q_{n+1}} = \frac{a_{n+1}p_n + p_{n-1}}{a_{n+1}q_n + q_{n-1}}$$

From the continued fraction expansion of $\alpha = \sqrt{N}$, namely

$$\alpha = \sqrt{N} = a_0 + \frac{1}{a_1+}\ \frac{1}{a_2+} \cdots \frac{1}{a_n+}\ \frac{1}{2a_0+} \cdots$$

if we replace the $(n+1)$-th partial quotient $a_{n+1} (= 2a_0)$ in the expression for the $(n+1)$-th convergent by the $(n+1)$-th complete quotient α_{n+1} we shall get the original number, namely $\alpha = \sqrt{N}$. So

$$\sqrt{N} = \frac{\alpha_{n+1}p_n + p_{n-1}}{\alpha_{n+1}q_n + q_{n-1}}.$$

Now

$$\alpha_{n+1} = 2a_0 + \frac{1}{a_1+}\ \frac{1}{a_2+} \cdots = a_0 + \sqrt{N}.$$

Substituting $\alpha_{n+1} = a_0 + \sqrt{N}$ in the above expression for $\sqrt{N}$ we get

$$\sqrt{N} = \frac{(a_0 + \sqrt{N})p_n + p_{n-1}}{(a_0 + \sqrt{N})q_n + q_{n-1}}.$$

On simplification this gives

$$Nq_n + (a_0q_n + q_{n-1})\sqrt{N} = a_0p_n + p_{n-1} + p_n\sqrt{N}.$$

Equating the rational and irrational parts we get

$$Nq_n = a_0p_n + p_{n-1} \qquad \text{i.e.} \qquad p_{n-1} = Nq_n - a_0p_n$$

$$a_0q_n + q_{n-1} = p_n \qquad \text{i.e.} \qquad q_{n-1} = p_n - a_0q_n$$

substituting these values of p_{n-1} and q_{n-1} in the relation

$$p_nq_{n-1} - p_{n-1}q_n = (-1)^{n-1}$$

we get

$$p_n(p_n - a_0 q_n) - (Nq_n - a_0 p_n)q_n = (-1)^{n-1}$$

That is,

$$p_n^2 - Nq_n^2 = (-1)^{n-1}.$$

So if the period $n+1$ of the continued fraction of $\sqrt{N}$ is even the numerator and denominator of the n-the convergent, namely p_n and q_n of $\sqrt{N}$ provide a solution of the Pell's equation $x^2 - Ny^2 = 1$. Thus a solution in this case is given by $x = p_n$, $y = q_n$. If the period $n+1$ of the continued fraction of $\sqrt{N}$ is odd then $(p_n^2 - Nq_n^2)^2 = (-1)^2 = 1$, that is $(p_n^2 + Nq_n^2)^2 - N(2p_n q_n)^2 = 1$ and so a solution to the Pell's equation $x^2 - Ny^2 = 1$ is given by

$$x = p_n^2 + Nq_n^2, \qquad y = 2p_n q_n$$

In each case the solution is "minimal" (or "least") in the sense described earlier. So we have the following algorithm for finding the minimal solution to the Pell's equation $x^2 - Ny^2 = 1$.

Step 1. Expand $\sqrt{N}$ into a continued fraction until the end of the first period, that is, until we get partial quotient which is equal to $2a_0 = 2[\sqrt{N}]$.

$$\begin{aligned}\sqrt{N} &= a_0 + \frac{1}{a_1+}\,\frac{1}{a_2+}\cdots\frac{1}{a_n+}\,\frac{1}{2a_0+}\cdots \\ &= (a_0; \overline{a_1, a_2, \cdots, a_n, 2a_0})\end{aligned}$$

So the period of the continued fraction is $n+1$.

Step 2. Compute the numerators and denominators (i.e., p_n and q_n) of the n-th convergent for the continued fraction of $\sqrt{N}$.

Step 3. If the period $n+1$ is even then the minimal solution of the Pell's equation is given by $x = p_n$, $y = q_n$. If the period $n+1$ is odd then $p_n^2 - Nq_n^2 = -1$ and the minimal solution to the Pell's equation $x^2 - Ny^2 = 1$ is obtained by squaring $p_n^2 - Nq_n^2 = -1$. Thus in this case $x = p_n^2 + Nq_n^2$, $y = 2p_n q_n$.

Two Examples

(1) Let $N = 31$. So we wish to solve the Pell's equation $x^2 - 31y^2 = 1$. As can be easily checked the continued fraction of $\sqrt{31}$ is

$$\begin{aligned}\sqrt{31} &= 5+\frac{1}{1+}\ \frac{1}{1+}\ \frac{1}{3+}\ \frac{1}{5+}\ \frac{1}{3+}\ \frac{1}{1+}\ \frac{1}{1+}\ \frac{1}{10+}\cdots \\ &= (5;\overline{1,1,3,5,3,1,1,10}).\end{aligned}$$

The period of this continued fraction is 8 and first eight partial quotients are $a_0 = 5, a_1 = a_2 = 1, a_3 = 3, a_4 = 5, a_5 = 3, a_6 = a_7 = 1$. We compute the values of $p_n's$ and $q_n's$ with the help of the following table:

i	-2	-1	0	1	2	3	4	5	6	7
a_i			5	1	1	3	5	3	1	1
p_i	0	1	5	6	11	39	206	657	863	1520
q_i	1	0	1	1	2	7	37	118	155	273.

Since the period of the continued fraction is even, we have $x = 1520$, $y = 273$ as the minimal solution to the Pell's equation $x^2 - 31y^2 = 1$. Incidentally, in this case, the Diophantine equation $x^2 - 31y^2 = -1$ has no solution, which is obvious since $\left(\frac{-1}{31}\right) = -1$, that is -1 is a quadratic nonresidue modulo 31. In fact, if $N > 1$ is square-free and its has a prime divisor $p \equiv 3 \pmod 4$, then the continued fraction of $\sqrt{N}$ has even period.

(2) Let us solve the Pell's equation $x^2 - 29y^2 = 1$. Here $N = 29$, and the continued fraction for $\sqrt{29}$ is

$$\begin{aligned}\sqrt{29} &= 5+\frac{1}{2+}\ \frac{1}{1+}\ \frac{1}{2+}\ \frac{1}{2+}\ \frac{1}{10+}\cdots \\ &= (5;\overline{2,1,1,2,10})\end{aligned}$$

This continued fraction has period 5. The partial quotients are $a_0 = 5,\ \ a_1 = 2,\ \ a_2 = a_3 = 1,\ \ a_4 = 2$. We compute the values of $p_n's$ and $q_n's$ with the help of the following table:

i	-2	-1	0	1	2	3	4
a_i			5	2	1	1	2
p_i	0	1	5	11	16	27	70
q_i	1	0	1	2	3	5	13

In this case, $p_4 = 70$, $q_4 = 13$ give the solution for the Diophantine equation $x^2 - 29y^2 = -1$. So the solution for Pell's equation $x^2 - 29y^2 = 1$ are given by $x = p_4^2 + 29 \cdot q_4^2 = 70^2 + 29 \cdot 13^2 = 9801$, and $y = 2p_4 \cdot q_4 = 2 \cdot 70 \cdot 13 = 1820$.

5. Chakravāla "Algorithm" For Finding a Solution of Pell's Equation

In this section we shall discuss what is known as *Chakravāla* "Algorithm" for finding a solution of Pell's equation which was developed by Indian mathematicians between 6th and 12th century. This method is attributed to *Brahamgupta* (born 598 AD) and *Bhascara Acharya* (born 1114 AD). These Indian mathematicians were convinced that their method works in all cases but did not supply proofs for the same. The form of the algorithm given below is a minor variation of the algorithm discussed in Vardarajan (1997).

Given an integer $N > 1$, which is not a perfect square, this algorithm determines the "smallest" pair (X, Y) of natural numbers X and Y satisfying $X^2 - NY^2 = 1$.

- Step 1. Let $s = [\sqrt{N}]$. If $s^2 + s > N$ then let $p_0 = s$ otherwise let $p_0 = s + 1$. Also let $q_0 = 1$ and $m_0 = p_0^2 - N$. That is to begin with we have

$$p_0 = [\sqrt{N}] + \left[\frac{N}{[\sqrt{N}] \cdot ([\sqrt{N}] + 1)}\right], \qquad q_0 = 1, \qquad m_0 = p_0^2 - N.$$

  Let a be the smallest positive integer satisfying $a \equiv -p_0 \pmod{|m_0|}$ and $a < \sqrt{N} + |m_0|$ (i.e., start with $-p_0$ and keep on adding $|m_0|$ until you get a positive number satisfying the condition $a < \sqrt{N} + |m_0|$). If $|a^2 - N| < (a + |m_0|)^2 - N$, then $x_0 = a$ otherwise $x_0 = a + |m_0|$. If $m_0 = 1$, then $(X, Y) = (p_0, q_0)$ is a solution. If not then go to the next step.

- Step 2. Given p_i, q_i, m_i, and x_i find $p_{i+1}, q_{i+1}, m_{i+1}$ as follows

$$p_{i+1} = \frac{p_i x_i + N q_i}{|m_i|}, \qquad q_{i+1} = \frac{p_i + q_i x_i}{|m_i|}, \qquad m_{i+1} = \frac{x_i^2 - N}{m_i}.$$

  Note that

$$p_{i+1}^2 - N q_{i+1}^2 = \left(\frac{p_i x_i + N q_i}{|m_i|}\right)^2 - N\left(\frac{p_i + q_i x_i}{|m_i|}\right)^2 = \frac{x_i^2 - N}{m_i} = m_{i+1}.$$

  So if $m_{i+1} = 1$ then $(X, Y) = (p_{i+1}, q_{i+1})$ and stop otherwise go to the next step.

- Step 3. Let $m = m_{i+1}$. Let a be the smallest positive integer satisfying the conditions $a \equiv -x_i \pmod{|m|}$ and $a < \sqrt{N} < a + |m|$. If $|a^2 - N| < (a + |m|)^2 - N$, then $x_{i+1} = a$ otherwise $x_{i+1} = a + |m|$. This is same as saying that
$$x_{i+1} = a + |m| \cdot \left[\frac{2N}{(a + |m|)^2 + a^2} \right].$$
Replace p_i, q_i, m_i, and x_i by $p_{i+1}, q_{i+1}, m_{i+1}$, and x_{i+1}. Go to step 2.

Chakravāla "Algorithm"

Solution of $X^2 - NY^2 = 1$ for $N = 13$

i	$\mathrm{p_i}$	$\mathrm{q_i}$	$\mathrm{m_i}$	$\mathrm{x_i}$
0	4	1	3	2
1	7	2	−3	4
2	18	5	−1	4
3	137	38	−3	2
4	256	71	3	4
5	**649**	**180**	1	

The solution of $X^2 - NY^2 = 1$ for $N = 61$.

i	$\mathrm{p_i}$	$\mathrm{q_i}$	$\mathrm{m_i}$	$\mathrm{x_i}$
0	8	1	3	7
1	39	5	−4	9
2	164	21	−5	6
3	453	58	5	9
4	1523	195	4	7
5	5639	722	−3	8
6	29718	3805	−1	8
7	469849	60158	−3	7
8	2319527	296985	4	9
9	9747957	1248098	5	6
10	26924344	3447309	−5	9
11	90520989	11590025	−4	7
12	335159612	42912791	3	8
13	**1766319049**	**226153980**	1	

The solution of $X^2 - NY^2 = 1$ for $N = 67$.

i	p_i	q_i	m_i	x_i
0	8	1	−3	7
1	41	5	6	5
2	90	11	−7	9
3	221	27	−2	9
4	1899	232	−7	5
5	3577	437	6	7
6	9053	1106	−3	8
7	**48842**	**5967**	1	

The solution of $X^2 - NY^2 = 1$ for $N = 97$.

i	p_i	q_i	m_i	x_i
0	10	1	3	11
1	69	7	8	5
2	128	13	−9	13
3	325	33	−8	11
4	847	86	−3	10
5	5604	569	−1	10
6	111233	11294	−3	11
7	773027	78489	−8	5
8	1434821	145684	9	13
9	3642669	369857	8	11
10	9493186	963887	3	10
11	**62809633**	**6377352**	1	

The solution of $X^2 - NY^2 = 1$ for $N = 106$.

i	$\mathbf{p_i}$	$\mathbf{q_i}$	$\mathbf{m_i}$	$\mathbf{x_i}$
0	10	1	−6	8
1	31	3	7	13
2	103	10	9	5
3	175	17	−9	13
4	453	44	−7	8
5	1184	115	6	10
6	4005	389	−1	10
7	81284	7895	6	8
8	247857	24074	−7	13
9	824855	80117	−9	5
10	1401853	136160	9	13
11	3628561	352437	7	8
12	9483830	921151	−6	10
12	**32080051**	**3115890**	1	

Edwards (1977), Vardarajan (1997) and Weil (1984) have detailed discussion of *Chakravāla* "Algorithm"

CHAPTER 9

Algebraic Number Theory

In this chapter, we introduce rudiments of elementary algebraic number theory. Using the ideas developed here, we give a proof of **Fermat's Last Theorem** for exponent three—namely, the Diophantine equation $x^3 + y^3 = z^3$ has no nontrivial solution.

In Appendix 1, we define algebraic numbers and algebraic integers and prove that the set of algebraic numbers form a field and set of algebraic integers form a ring. We first describe the algebraic integers and units in quadratic number fields, namely, the fields $\mathbb{Q}(\sqrt{m}) = \{a + b\sqrt{m} : a, b \in \mathbb{Q}\}$, where m is a square-free integer. Evidently, the elements of the form $u + v\sqrt{m}$ where $u, v \in \mathbb{Z}$ are algebraic integers. In fact, these are all the algebraic integers in $\mathbb{Q}(\sqrt{m})$ if $m \equiv 2$ or $3 \pmod 4$. When $m \equiv 1 \pmod 4$, the set of algebraic integers in $\mathbb{Q}(\sqrt{m})$ consists of the elements of the form $\dfrac{u + v\sqrt{m}}{2}$ where u and v in $\mathbb{Z}$ have the same parity (i.e., either both are odd or both are even) as we soon demonstrate.

In what follows $m \neq 1$ will denote a (positive or negative) square-free integer. We can express an arbitrary element $\alpha = a+b\sqrt{m}$ of $\mathbb{Q}(\sqrt{m})$ in the form $\dfrac{u + v\sqrt{m}}{w}$ where $u, v, w \in \mathbb{Z}$, $w > 0$, and $gcd(u, v, w) = 1$. The minimal polynomial of the algebraic number $\alpha = \dfrac{u + v\sqrt{m}}{w}$ over the field of the rational numbers is

$$x^2 - \frac{2u}{w}x + \frac{u^2 - v^2m}{w^2}.$$

In order that α is an algebraic integer, the coefficients $\dfrac{2u}{w}$ and $\dfrac{u^2 - v^2m}{w^2}$ of the above polynomial must be (rational) integers. That is, w divides $2u$ and w^2 divides $(u^2 - v^2m)$. There are two cases to be discussed depending upon whether w is odd or even.

Case I. w is odd.

In this case $w|u$, and therefore $w^2|v^2m$. Since $gcd(u, v, w) = 1$ and $w|u$, we have $gcd(w, v) = 1$ and so $w^2|m$. Since m is square-free integer, we must have $w = 1$. So $\alpha = u + v\sqrt{m}$ where $u, v \in \mathbb{Z}$.

Case II. w is even.

Let $w = 2t$. So $t|u$ and $4t^2|(u^2 - v^2m)$. It follows that $t^2|v^2m$. Since $gcd(u, v, w) = 1$, it follows that $t = 1$. Therefore $w = 2$. Again, since m is square-free integer and $4|(u^2 - v^2m)$, u and v have the same parity. Since $w = 2$, and $gcd(u, v, w) = 1$, u and v are odd and since $4|(u^2 - v^2m)$, we have $m \equiv 1 \pmod 4$.

We summarize what we have shown above in the following theorem.

Theorem: Let m be a square-free rational integer. The algebraic integers in $\mathbb{Q}(\sqrt{m})$ are

$$\{u + v\sqrt{m} : u, v \in \mathbb{Z}\} \quad \text{if} \quad m \equiv 2, \text{ or } 3 \pmod 4,$$

$$\left\{\frac{u + v\sqrt{m}}{2} : u, v \in \mathbb{Z}, u \equiv v \bmod 2\right\} \quad \text{if} \quad m \equiv 1 \pmod 4.$$

Units in Quadratic Fields

An algebraic integer α is said to be a *unit* if there exists an algebraic integer β such that $\alpha\beta = 1$. Evidently, the set of all units forms a multiplicative group. In order to describe the units in a quadratic field, it is useful to introduce the notion of a norm in $\mathbb{Q}(\sqrt{m})$. For $a + b\sqrt{m}$ in $\mathbb{Q}(\sqrt{m})$, we define

$$N(a + b\sqrt{m}) = (a + b\sqrt{m})(a - b\sqrt{m}) = a^2 - b^2m.$$

Evidently, this norm function is multiplicative, that is, $N(\alpha\beta) = N(\alpha)N(\beta)$, for all $\alpha, \beta \in \mathbb{Q}(\sqrt{m})$. It follows that an algebraic integer $\alpha \in \mathbb{Q}(\sqrt{m})$ is a unit if and only if $N(\alpha) = \pm 1$. A unit by definition is an algebraic integer. We add that there are algebraic numbers of norm one which are not algebraic integers. For example, the algebraic number $\dfrac{19 + 6\sqrt{2}}{17} \in \mathbb{Q}(\sqrt{2})$ is not an algebraic integer and

$$N\left(\frac{19 + 6\sqrt{2}}{17}\right) = \left(\frac{19 + 6\sqrt{2}}{17}\right) \times \left(\frac{19 - 6\sqrt{2}}{17}\right) = \frac{361 - 72}{289} = 1.$$

Units in Complex Quadratic Fields

We first discuss the case of imaginary quadratic fields. Suppose that m is negative. First suppose that $m < -4$. If $m \equiv 2, 3 \pmod 4$, an algebraic integer $\alpha \in \mathbb{Q}(\sqrt{m})$ has the form $u + v\sqrt{m}$ where $u, v \in \mathbb{Z}$. Now $N(u + v\sqrt{m}) = u^2 - v^2m > 0$ (since $m < 0$). Therefore, in order that $u + v\sqrt{m}$ is a unit in $\mathbb{Q}(\sqrt{m})$, we must have $u^2 - mv^2 = 1$. Since $m < -4$, the only solutions of this Diophantine equation are $u = \pm 1, v = 0$. This corresponds to the units ± 1.

Now suppose that $m \equiv 1 \pmod 4$, and we are still assuming that $m < -4$. In this case, an algebraic integer $\alpha \in \mathbb{Q}(\sqrt{m})$ is an element of the form $\alpha = \dfrac{u + v\sqrt{m}}{2}$, where $u, v \in \mathbb{Z}$ are both even or both odd. So α will be a unit if and only if

$$N(\alpha) = N\left(\frac{u + v\sqrt{m}}{2}\right) = \frac{u^2 - mv^2}{4} = 1,$$

that is, if and only if $u^2 - mv^2 = 4$. Again since $m < -4$, the only solutions to the above Diophantine equation are $u = \pm 2$ and $v = 0$ and thus $\alpha = \pm 1$.

We will deal with the remaining case of negative values of m (i.e., $m = -1, -2$, and -3) one at a time.

If $m = -1$ then we write $\mathbb{Q}(\sqrt{-1}) = \mathbb{Q}(i)$. A unit $\alpha \in \mathbb{Z}[i] \subset \mathbb{Q}(i)$ is of the form $u + vi$, where $u, v \in \mathbb{Z}$. We have $N(\alpha) = u^2 + v^2 = 1$. The only integral solutions of $u^2 + v^2 = 1$ are given by $u = \pm 1, v = 0$ and $u = 0, v = \pm 1$. This corresponds to the four units, namely, $1, -1, i, -i$.

If $m = -2$ then a unit $u + v\sqrt{-2} \in \mathbb{Z}[\sqrt{-2}] \subset \mathbb{Q}(\sqrt{-2})$ must satisfy $u^2 + 2v^2 = 1$. The only integral solutions of $u^2 + 2v^2 = 1$ are given by $u = \pm 1, v = 0$. This corresponds to the two units, namely, 1 and -1.

Finally if $m = -3$, a unit $\dfrac{u + v\sqrt{-3}}{2}$, where $u \equiv v \pmod 2$, in $\mathbb{Q}(\sqrt{-3})$ satisfies $u^2 + 3v^2 = 4$. The solutions of this Diophantine equation are by $u = \pm 2, v = 0$ and $u = \pm 1, v = \pm 1$. This corresponds to the six units $\pm 1, \pm\omega, \pm\omega^2$, where $\omega = \dfrac{-1 + \sqrt{-3}}{2}$ is primitive cube root of unity with positive imaginary part. We note that the six units in $\mathbb{Q}(\sqrt{-3})$ form a cyclic group generated by $\dfrac{1 - \sqrt{-3}}{2} = -\omega$. The following table gives a summary of what we have proved above about the units in the ring of algebraic integers of the complex quadratic field $\mathbb{Q}(m)$, where $m < 0$ is a square-free integer.

m	Units in $\mathbb{Q}(\sqrt{m})$
-1	$\pm 1, \pm i$
-3	$\pm 1, \pm\omega, \pm\omega^2$
$m = -2$ or $m < -3$	± 1

Units in Real Quadratic Fields

We now describe the units in the ring of algebraic integers of the quadratic field $\mathbb{Q}(\sqrt{m})$, where $m > 1$ is a square-free integer. We show that in case of real quadratic fields $\mathbb{Q}(\sqrt{m})$, the group of units is isomorphic to direct product of group of order 2 and an infinite cyclic group. First, we show that for any $c > 1$, there are only finitely many units of $\mathbb{Q}(\sqrt{m})$ lying in the interval $(1, c)$. We consider the two cases $m \equiv 2, 3 \pmod 4$ and $m \equiv 1 \pmod 4$ separately.

Case I $m \equiv 2, 3 \pmod 4$. In this case, the set of integers in $\mathbb{Q}(\sqrt{m})$ is

$$\{a + b\sqrt{m} : a, b \in \mathbb{Z}\}.$$

We proved in Chapter 8 that the Diophantine equation $x^2 - my^2 = 1$ has infinitely many solutions. We call $(a, b) \in \mathbb{Z} \times \mathbb{Z}$ a solution of $x^2 - my^2 = 1$ if $a^2 - mb^2 = 1$. Clearly, if $(a, b) \in \mathbb{Z} \times \mathbb{Z}$ is a solution of $x^2 - my^2 = 1$ then so is $(\pm a, \pm b)$. So we can pick up a solution $(a, b) \in \mathbb{Z} \times \mathbb{Z}$ of $x^2 - my^2 = 1$ such that $a > 0$ and $b > 0$. Thus, $\mathbb{Q}(\sqrt{m})$ has a unit $\alpha = a + b\sqrt{m}$ such that $a, b > 0$. For any unit $\alpha = a + b\sqrt{m}$, $N(\alpha) = a^2 - b^2m = \pm 1$. We wish to show that for any $c > 1$ there are finitely many units of $\mathbb{Q}(\sqrt{m})$ in the interval $(1, c)$. If there are no units of $\mathbb{Q}(\sqrt{m})$ in the interval $(1, c)$ then we have nothing to prove. Let $\alpha = a + b\sqrt{m} \in \mathbb{Z}[\sqrt{m}] \subset \mathbb{Q}(\sqrt{m})$ be a unit such that $1 < \alpha < c$. We note that

$$\{\pm\alpha, \pm\frac{1}{\alpha}\} = \{\pm a \pm b\sqrt{m}\}.$$

So the largest of the four units $\pm\alpha, \pm\frac{1}{\alpha}$ has positive rational and irrational parts and is the only unit among these four that exceeds 1. So $\alpha = a + b\sqrt{m}$, $a, b > 0$, $N(\alpha) = a^2 - b^2m = \pm 1$. The minimal polynomial of $\alpha = a + b\sqrt{m}$ is $x^2 - 2ax \pm 1$, or simply $x^2 + Ax \pm 1$, where $A = -2a \in \mathbb{Z}$. Since $|A| = 2a < 2(a + b\sqrt{m}) < 2c$, there are finitely many choices for A. So there are finitely units of $\mathbb{Q}(\sqrt{m})$ lying in the interval $(1, c)$.

Case II $m \equiv 1 \pmod 4$. In this case, the set of integers in $\mathbb{Q}(\sqrt{m})$ is

$$\left\{\frac{a + b\sqrt{m}}{2} : a, b \in \mathbb{Z}, a \equiv b \pmod 2\right\}.$$

Since the Diophantine equation $x^2 - my^2 = 1$ has infinitely many solutions then so does the equation $x^2 - my^2 = 4$ and in the latter case x and y must have same parity since m is odd. It follows that $\mathbb{Q}(\sqrt{m})$ has infinitely many units. So if $\alpha = \dfrac{a + b\sqrt{m}}{2} \neq \pm 1$ is a unit then among the four units $\dfrac{\pm a \pm b\sqrt{m}}{2}$ the largest one is the only one that exceeds 1 and it has positive rational and irrational parts. Let $c > 1$. If there are no units of $\mathbb{Q}(\sqrt{m})$ in $(1, c)$ then we have nothing to prove. Let $\alpha = \dfrac{a + b\sqrt{m}}{2} \in (1, c)$ be a unit of $\mathbb{Q}(\sqrt{m})$. So $a, b > 0$ and $N(\alpha) = \dfrac{a^2 - b^2 m}{4} = \pm 1$. The minimal polynomial of α is $x^2 - ax \pm 1$. We have $\dfrac{a}{2} < \alpha < c$ and as before we conclude that there are only finitely many possibilities for a and therefore the same is true for units α in the interval $(1, c)$.

We now show that the group of units in the real quadratic field is isomorphic to direct product of an infinite cyclic group and a group of order two.

Theorem: Let $m > 1$ be a square-free integer. Then the group G of the units in the ring of algebraic integers of the quadratic field $\mathbb{Q}(\sqrt{m})$ is isomorphic to the direct product of a group of order two and an infinite cyclic group.

Proof: We note that if α is a unit in $\mathbb{Q}(\sqrt{m})$ then so are $\pm\alpha$ and $\pm\alpha^{-1}$. Evidently, if $\alpha \neq \pm 1$ exactly one of these units is greater than one. So the group G has positive units, that is, the units which are positive (real) numbers. We show that the positive elements in G constitute an infinite cyclic group. By previous discussion, one can find $\gamma \in G$, such that

$$\gamma = Min\,\{\alpha : \alpha > 1, \alpha \in G\}.$$

We show that the set of positive elements in G is precisely $\langle\gamma\rangle$, the cyclic group generated by γ. It is sufficient to show that ever $\alpha \in G$, $\alpha > \gamma$ is a positive integral power of γ. Let $\alpha \in G$ and $\alpha > 1$. If α is not an exact integral power of γ then there exists a unique integer m such that $\gamma^m < \alpha < \gamma^{m+1}$. This gives $1 < \dfrac{\alpha}{\gamma^m} < \gamma$. Since $\dfrac{\alpha}{\gamma^m}$ is a unit, this contradicts the choice of γ. Therefore, every element $\alpha > 1$ of G is a positive integral power of γ. This in turn implies that every positive element $\alpha \in G$ is an integral power (positive or negative) of γ. Now consider the map

$$\begin{array}{ccc} G & \longrightarrow & \{1, -1\} \times \langle\gamma\rangle \\ \alpha & \longrightarrow & \left(\dfrac{\alpha}{|\alpha|}, |\alpha|\right). \end{array}$$

This mapping is easily seen to be a homomorphism. Since every positive element of G is an integral power of γ, this mapping is onto. The kernel of this mapping is identity of G. Hence

$$G \cong \{1, -1\} \times \langle\gamma\rangle,$$

that is, G is isomorphic to the direct product of a group of order two and an infinite cyclic group.

Note: $\gamma = Min\{\alpha : \alpha > 1, \alpha \in G\}$ is called the *Fundamental Unit* of the quadratic field $\mathbb{Q}(\sqrt{m})$.

We mention here without proof that the above theorem is a special case of a famous theorem of Dirichlet which states that if α is an algebraic integer of degree n and the minimal polynomial of α has r_1 real zeros and $2r_2$ complex zeros then the group of units in $\mathbb{Q}(\alpha)$ is isomorphic to the direct product of the finite group U of units in $\mathbb{Q}(\alpha)$ lying on the unit circle with $r(= r_1 + r_2 - 1)$ copies of an infinite cyclic group. In the real quadratic case discussed above $U = \{1, -1\}$ and $r = 1$.

1. Euclidean Quadratic Fields

Let D be an integral domain. A function

$$\phi : D \longrightarrow \mathbb{N} \cup \{0\}$$

is called a norm on D. An integral domain D is said to be an Euclidean domain if there is a norm on D such that for any two elements $\alpha, \beta \in D$ with $\beta \neq 0$, there exist γ and δ in D such that $\alpha = \beta\gamma + \delta$, where $\delta = 0$ or $\phi(\delta) < \phi(\beta)$. We refer to such a function ϕ as Euclidean function or Euclidean norm on the integral domain D.

As we encounter integral domains other than $\mathbb{Z}$, we need to find out if an integral domain in question has, or does not have, Euclidean property. In a beginning course on Abstract, Algebra students are taught that

$$\text{Euclidean Domain} \Longrightarrow \text{Principal Ideal Domain} \Longrightarrow \text{Unique Factorization Domain}$$

so if we know that an integral domain D has Euclidean property then we know that it is a unique factorization domain and we can carry out arithmetic in D just as we do in $\mathbb{Z}$. Some examples of integral domains having Euclidean property are

- $\mathbb{Z}[i] \subset \mathbb{Q}(i)$.
- $\mathbb{Z}[\omega] = \mathbb{Z}[\frac{-1+\sqrt{-3}}{2}] \subset \mathbb{Q}(\sqrt{-3})$.
- $\mathbb{F}[x]$, where $\mathbb{F}$ is a field.

In this section, we study the ring of integers in some quadratic fields which have Euclidean property. If the ring of integers of the quadratic field $\mathbb{Q}(\sqrt{m})$, where m is a square-free integer, has Euclidean property then by abuse of language we say that $\mathbb{Q}(\sqrt{m})$ is Euclidean rather than using somewhat longer but more precise term namely that the ring of all algebraic integers in $\mathbb{Q}(\sqrt{m})$ is Euclidean. In the quadratic field $\mathbb{Q}(\sqrt{m})$, we define the norm function as follows:

$$\begin{aligned}\phi : \mathbb{Q}(\sqrt{m}) &\longrightarrow \mathbb{Q}\\ \phi(a+b\sqrt{m}) &= |N(a+b\sqrt{m}| = |(a+b\sqrt{m})(a-b\sqrt{m}| = |a^2-mb^2|.\end{aligned}$$

Therefore for $\alpha \in \mathbb{Q}(\sqrt{m})$, $\phi(\alpha)$ is the absolute value of the product of α and its algebraic conjugate. This function ϕ has the property that it is multiplicative, namely $\phi(\alpha\beta) = \phi(\alpha)\phi(\beta)$ for $\alpha, \beta \in \mathbb{Q}(\sqrt{m})$. We note that when restricted to ring of (algebraic) integers in $\mathbb{Q}(\sqrt{m})$, the function ϕ takes values in $\mathbb{N} \cup \{0\}$.

Earlier in this chapter, we discussed the ring of integers in the quadratic fields. It is known that $\mathbb{Q}(\sqrt{m})$ is Euclidean (with respect to the norm function N defined above) for the following values of m:

$$-11, -7, -3, -2, -1, 2, 3, 5, 6, 7, 11, 13, 17, 19, 21, 29, 33, 37, 41, 57, 73.$$

The complete proof the above statement is beyond the scope of present text. However, for certain values of m a simple proof can be given. That is what we do in the following theorem.

Theorem $\mathbb{Q}(\sqrt{m})$ is Euclidean with respect to the norm function ϕ defined above for $m = -11, -7, -3, -2, -1, 2, 3, 5, 13, 17$.

Proof The proof of this theorem is divided into four cases.

Case 1 $m = -2, -1, 2, 3$.

In this case $m \equiv 2, 3 \pmod 4$. Therefore, the ring of integers in $\mathbb{Q}(\sqrt{m})$ is $\mathbb{Z}[\sqrt{m}] = \{a+b\sqrt{m};\ \ a, b \in \mathbb{Z}\}$. Let $\alpha, \beta \in \mathbb{Z}[\sqrt{m}]$, $\beta \neq 0$. Suppose $\alpha = a+b\sqrt{m}$ and $\beta = c+d\sqrt{m}$. We have

$$\begin{aligned}\frac{\alpha}{\beta} &= \frac{a+b\sqrt{m}}{c+d\sqrt{m}}\\ &= \frac{(a+b\sqrt{m})(c-d\sqrt{m})}{c^2-md^2}\\ &= \frac{(ac-bdm)+(bc-ad)\sqrt{m}}{c^2-md^2}\\ &= \frac{ac-bdm}{c^2-md^2} + \frac{bc-ad}{c^2-md^2}\cdot\sqrt{m}\\ &= x+y\sqrt{m}\end{aligned}$$

where

$$x = \frac{ac - bdm}{c^2 - md^2} \in \mathbb{Q}, \qquad \text{and } y = \frac{bc - ad}{c^2 - md^2} \in \mathbb{Q}$$

Let $e, f \in \mathbb{Z}$ be such that

$$|x - e| \leq \frac{1}{2}, \quad |y - f| \leq \frac{1}{2}.$$

We have

$$\begin{aligned} \frac{\alpha}{\beta} &= x + y\sqrt{m} + e + f\sqrt{m} - (e + f\sqrt{m}) \\ &= e + f\sqrt{m} + (x - e) + (y - f)\sqrt{m} \\ \alpha &= \beta(e + f\sqrt{m}) + \{(x - e) + (y - f)\sqrt{m}\} \cdot \beta. \end{aligned}$$

We let $\gamma = e + f\sqrt{m}$ and $\delta = \{(x - e) + (y - f)\sqrt{m}\} \cdot \beta$. Then we have $\alpha = \beta\gamma + \delta$, where $\gamma, \delta \in \mathbb{Z}[\sqrt{m}]$, and

$$\begin{aligned} \phi(\delta) &= \phi(\{(x - e) + (y - f)\sqrt{m}\} \cdot \beta) \\ &= \phi(\{(x - e) + (y - f)\sqrt{m}\}) \cdot \phi(\beta) \\ &= |(x - e)^2 - (y - f)^2 m| \cdot \phi(\beta) \\ &\leq \left(\frac{1}{4} + \frac{|m|}{4}\right) \cdot \phi(\beta) \\ &< \phi(\beta) \qquad \text{if } m = -2, -1, 2. \end{aligned}$$

If $m = 3$ and $|x - e| < \frac{1}{2}$ or $|y - f| < \frac{1}{2}$ then $|(x - e)^2 - (y - f)^2 m| < \frac{1}{4} + \frac{3}{4} = 1$ and we have $\phi(\delta) < \phi(\beta)$. If $m = 3$ and $|x - e| = |y - f| = \frac{1}{2}$ then we have

$$\left|(x - e)^2 - (y - f)^2 m\right| = \left|\frac{1}{4} - \frac{3}{4}\right| = \frac{1}{2} < 1$$

and so in this case as well we have $\phi(\delta) < 1$.

Case 2 $m = -11, -7, -3, 5$. In this case, $m \equiv 1 \pmod 4$ and so the ring of integers in $\mathbb{Q}(\sqrt{m})$ is

$$\begin{aligned} D = \mathbb{Z}\left[\frac{1 + \sqrt{m}}{2}\right] &= \left\{u + v\left(\frac{1 + \sqrt{m}}{2}\right); \quad u, v \in \mathbb{Z}\right\} \\ &= \left\{\frac{a + b\sqrt{m}}{2}; a, b \in \mathbb{Z}, a \equiv b \pmod 2\right\}. \end{aligned}$$

Let $\alpha, \beta \in D$, $\beta \neq 0$ be

$$\alpha = \frac{a+b\sqrt{m}}{2}, \quad \beta = \frac{c+d\sqrt{m}}{2}, \quad a \equiv b \pmod 2, \qquad \text{and } c \equiv d \pmod 2.$$

As in case 1, we have

$$\begin{aligned} \frac{\alpha}{\beta} &= \frac{a+b\sqrt{m}}{c+d\sqrt{m}} \\ &= \frac{(ac-bdm)+(bc-ad)\sqrt{m}}{c^2-md^2} \\ &= \frac{ac-bdm}{c^2-md^2} + \frac{bc-ad}{c^2-md^2}\cdot\sqrt{m} \\ &= x+y\sqrt{m} \end{aligned}$$

where

$$x = \frac{ac-bdm}{c^2-md^2} \in \mathbb{Q}, \qquad \text{and } y = \frac{bc-ad}{c^2-md^2} \in \mathbb{Q}$$

Suppose that $r, s \in \mathbb{Z}$ be such that $r + s\left(\frac{1+\sqrt{m}}{2}\right)$ is the "quotient" when α is divided by β in D. Our purpose is to find such $r, s \in D$. We write

$$\begin{aligned} \frac{\alpha}{\beta} &= x+y\sqrt{m}+r+s\left(\frac{1+\sqrt{m}}{2}\right) - \left(r+s\left(\frac{1+\sqrt{m}}{2}\right)\right) \\ &= x+y\sqrt{m}+r+s\left(\frac{1+\sqrt{m}}{2}\right) - \left\{\frac{2r+s+s\sqrt{m}}{2}\right\} \\ &= r+s\left(\frac{1+\sqrt{m}}{2}\right) + x - \frac{2r+s}{2} + \left(y-\frac{s}{2}\right)\sqrt{m} \\ &= r+s\left(\frac{1+\sqrt{m}}{2}\right) + \left(\frac{2x-s}{2}-r\right) + \left(\frac{2y-s}{2}\right)\sqrt{m} \\ \alpha &= \left\{r+s\left(\frac{1+\sqrt{m}}{2}\right)\right\}\cdot\beta + \left\{\left(\frac{2x-s}{2}-r\right)+\left(\frac{2y-s}{2}\right)\sqrt{m}\right\}\cdot\beta \\ &= \gamma\beta+\delta \end{aligned}$$

where

$$\gamma = r+s\left(\frac{1+\sqrt{m}}{2}\right) \quad \text{and } \delta = \left\{\left(\frac{2x-s}{2}-r\right)+\left(\frac{2y-s}{2}\right)\sqrt{m}\right\}\beta.$$

First we choose $s \in \mathbb{Z}$ such that $|2y - s| \le \frac{1}{2}$. *After* having chosen s, we choose $r \in \mathbb{Z}$ such that $\left|\frac{2x-s}{2} - r\right| \le \frac{1}{2}$. Then we have

$$\begin{aligned}
\phi(\delta) &= \phi\left(\left\{\left(\frac{2x-s}{2} - r\right) + \left(\frac{2y-s}{2}\sqrt{m}\right)\right\}\cdot\beta\right)\\
&= \left|\left(\frac{2x-s}{2} - r\right)^2 - \left(\frac{2y-s}{2}\right)^2 m\right|\cdot\phi(\beta)\\
&\le \left(\frac{1}{4} + \frac{|m|}{16}\right)\cdot\phi(\beta)\\
&\le \left(\frac{1}{4} + \frac{11}{16}\right)\phi(\beta) < \phi(\beta).
\end{aligned}$$

Case 3 $m = 13$. We know that the integers in $\mathbb{Q}(\sqrt{13})$ are

$$\left\{\frac{a+b\sqrt{13}}{2} : a, b \in \mathbb{Z}, a \equiv b \pmod 2\right\} = \left\{u + \frac{1+\sqrt{13}}{2}v : u, v \in \mathbb{Z}\right\}.$$

Let α and $\beta \neq 0$ be two integers in $\mathbb{Q}(\sqrt{13})$ and $\frac{\alpha}{\beta} = x + y\sqrt{13}$, $x, y \in \mathbb{Q}$. For any integers $r, s \in \mathbb{Z}$ (yet to be chosen), we can write

$$\begin{aligned}
\frac{\alpha}{\beta} &= x + y\sqrt{13} = r + \frac{1+\sqrt{13}}{2}s + (x + y\sqrt{13}) - \left(r + \frac{1+\sqrt{13}}{2}s\right)\\
&= r + \frac{1+\sqrt{13}}{2}s + \left(x - r - \frac{s}{2}\right) + \left(y - \frac{s}{2}\right)\sqrt{13}\\
&= r + \frac{1+\sqrt{13}}{2}s + \left(x - r - \frac{s}{2}\right) + \left(\frac{2y-s}{2}\right)\sqrt{13}.
\end{aligned}$$

We choose integers $r, s \in \mathbb{Z}$ as follows. We first choose $s \in \mathbb{Z}$ such that $|2y - s| \le \frac{1}{2}$. Having chosen s, we chose $r \in \mathbb{Z}$ such that $\left|x - \frac{s}{2} - r\right| \le \frac{1}{2}$. Then

$$\phi\left\{\left(x - \frac{s}{2} - r\right) + \left(\frac{2y-s}{2}\right)\sqrt{13}\right\} = \left|\left(x - \frac{s}{2} - r\right)^2 - \left(\frac{2y-s}{2}\right)^2 13\right|.$$

Finally

$$\left|\left(x - \frac{s}{2} - r\right)^2 - \left(\frac{2y-s}{2}\right)^2 13\right| \le \text{Max}\left\{\left(x - \frac{s}{2} - r\right)^2, \left(\frac{2y-s}{2}\right)^2 13\right\} \le \frac{13}{16} < 1.$$

We have $\alpha = \beta \cdot \gamma + \delta$, where $q = r + \frac{1+\sqrt{13}}{2}s$, $\delta = \alpha - \beta \cdot \gamma \in \mathbb{Z}\left[\frac{1+\sqrt{13}}{2}\right]$. Now

$$\delta = \beta\left\{\left(x - \frac{s}{2} - r\right) + \left(\frac{2y-s}{2}\right)\sqrt{13}\right\}, \quad \phi(\delta) < \phi(\beta).$$

Case 4 $m = 17$. We know that the integers in $\mathbb{Q}(\sqrt{17})$ are

$$\left\{\frac{a+b\sqrt{17}}{2} : a, b \in \mathbb{Z}, a \equiv b \pmod 2\right\} = \left\{u + \frac{1+\sqrt{17}}{2}v : u, v \in \mathbb{Z}\right\}.$$

Let α and $\beta \neq 0$ be two integers in $\mathbb{Q}(\sqrt{17})$ and $\frac{\alpha}{\beta} = x + y\sqrt{17}$, $x, y \in \mathbb{Q}$. For any integers $r, s \in \mathbb{Z}$ (yet to be chosen), we can write

$$\begin{aligned}
\frac{\alpha}{\beta} &= x + y\sqrt{17} = r + \frac{1+\sqrt{17}}{2}s + (x + y\sqrt{17}) - \left(r + \frac{1+\sqrt{17}}{2}s\right) \\
&= r + \frac{1+\sqrt{17}}{2}s + \left(x - r - \frac{s}{2}\right) + \left(y - \frac{s}{2}\right)\sqrt{17} \\
&= r + \frac{1+\sqrt{17}}{2}s + \left(x - r - \frac{s}{2}\right) + \left(\frac{2y-s}{2}\right)\sqrt{17}
\end{aligned}.$$

We choose integers $r, s \in \mathbb{Z}$ as follows. We first choose $s \in \mathbb{Z}$ such that $|2y - s| \leq \frac{1}{2}$. Having chosen s, we chose $r \in \mathbb{Z}$ as follows:

- If $2y - s = 0$ then choose r such that $\left|x - r - \frac{s}{2}\right| \leq \frac{1}{2}$.
- If $2y - s \neq 0$ but $x - \frac{s}{2} \in \mathbb{Z}$ then choose $r = x - \frac{s}{2} + 1$.
- If $2y - s \neq 0$, and $x - \frac{s}{2} \notin \mathbb{Z}$ then choose r such that $\frac{1}{2} \leq \left|x - r - \frac{s}{2}\right| < 1$.

We have $\alpha = \beta \cdot \gamma + \delta$ where $\gamma = r + \frac{1+\sqrt{17}}{2}s$, and

$$\delta = \alpha - \beta \cdot \gamma = \beta \cdot \left\{\left(x - r - \frac{s}{2}\right) + \left(\frac{2y-s}{2}\right)\sqrt{17}\right\}.$$

To show that $\phi(\delta) < \phi(\beta)$, we only need to prove the following:

$$\left|\left(x - r - \frac{s}{2}\right)^2 - 17\left(\frac{2y-s}{2}\right)^2\right| < 1.$$

- If $2y - s = 0$ then

$$\left|\left(x - r - \frac{s}{2}\right)^2 - 17\left(\frac{2y-s}{2}\right)^2\right| = \left(x - r - \frac{s}{2}\right)^2 \leq \frac{1}{4} < 1.$$

- If $2y - s \neq 0$, and $x - \frac{s}{2} \in \mathbb{Z}$ then $\left(x - \frac{s}{2} - r\right)^2 = 1$ and therefore
$$\left|\left(x - r - \frac{s}{2}\right)^2 - \left(\frac{2y-s}{2}\right)^2 17\right| = \left|\left(\frac{2y-s}{2}\right)^2 17 - 1\right| < 1.$$
- If $2y - s \neq 0$, and $x - \frac{s}{2} \notin \mathbb{Z}$ then $\frac{1}{2} \leq \left|x - r - \frac{s}{2}\right| < 1$ and so $\frac{1}{4} \leq \left(x - \frac{s}{2} - r\right)^2 < 1$.
If $\left(x - r - \frac{s}{2}\right)^2 > 17\left(\frac{2y-s}{2}\right)^2$ then we have
$$\left|\left(x - r - \frac{s}{2}\right)^2 - 17\left(\frac{2y-s}{2}\right)^2\right| < \left(x - r - \frac{s}{2}\right)^2 < 1.$$
However, if $\left(x - r - \frac{s}{2}\right)^2 \leq 17\left(\frac{2y-s}{2}\right)^2$ then
$$\left|\left(x - r - \frac{s}{2}\right)^2 - 17\left(\frac{2y-s}{2}\right)^2\right| = \left(\frac{2y-s}{2}\right)^2 17 - \left(x - r - \frac{s}{2}\right)^2 \leq 17\left(\frac{2y-s}{2}\right)^2$$
$$-\frac{1}{4} \leq \frac{17}{16} - \frac{1}{4} = \frac{13}{16} < 1.$$

This shows that $\mathbb{Q}(\sqrt{17})$ is Euclidean and completes the proof of the theorem.

Note The integral domain consisting of all algebraic integers in the quadratic fields $\mathbb{Q}(\sqrt{14})$ and $\mathbb{Q}(\sqrt{29})$ are also known to be Euclidean, but with respect to some norm function different from the one discussed in the previous theorem. See Alaca and Williams (2004) and Harper (2004).

2. Fermat's Last Theorem for Exponent 3

In this chapter, one of our objective is to give a proof of Fermat's last theorem for exponent 3. Therefore, from now on in this chapter, we confine our attention to the field $\mathbb{K} = \mathbb{Q}(\sqrt{-3}) = \mathbb{Q}(\omega)$, where $\omega = \frac{-1+\sqrt{-3}}{2}$. We have seen that the ring of algebraic integers in this field is the set
$$\left\{\frac{a+b\sqrt{-3}}{2} : a, b \in \mathbb{Z}, a \equiv b \pmod{2}\right\} = \{a + b\omega, a, b \in \mathbb{Z}\}.$$
From our discussion in the previous section, we know that $\mathbb{Z}[\omega]$ is a unique factorization domain. We have also seen that the units in $\mathbb{Z}[\omega]$ are $\pm 1, \pm\omega$, and $\pm\omega^2$. Also to recapitulate the map
$$\begin{aligned} N &: \mathbb{Z}[\omega] \longrightarrow \mathbb{N} \cup \{0\} \\ N(a + b\omega) &= a^2 - ab + b^2 \end{aligned}$$

has the following properties:

1. For $\alpha \in \mathbb{Z}[\omega]$, $\quad N(\alpha) \geq 0$. $N(\alpha) = 0$ if and only if $\alpha = 0$.
2. N is multiplicative. That is for $\alpha, \beta \in \mathbb{Z}[\omega]$, $N(\alpha\beta) = N(\alpha)N(\beta)$.
3. For $\alpha, \beta \in \mathbb{Z}[\omega]$, β divides α implies that $N(\beta)$ divides $N(\alpha)$.
4. Given α, β in $\mathbb{Z}[\omega]$, $\beta \neq 0$, there exists γ, δ in $\mathbb{Z}[\omega]$ such that $\alpha = \beta\gamma + \delta$, where $0 \leq N(\delta) < N(\beta)$.

Lemma 1

If $\alpha \in \mathbb{Z}[\omega]$ is such that $N(\alpha)$ is a rational prime p (i.e., a prime in $\mathbb{Z}$) then α is a prime in $\mathbb{Z}[\omega]$.

Proof

If β divides α then we can write $\alpha = \beta \cdot \gamma$ for some $\gamma \in \mathbb{Z}[\omega]$. Then we have $p = N(\alpha) = N(\beta) \cdot N(\gamma)$. Therefore, either $N(\beta) = 1$ or $N(\gamma) = 1$ and so β or γ is a unit and α is a prime in $\mathbb{Z}[\omega]$.

Lemma 2

Given a prime $\pi \in \mathbb{Z}[\omega]$ there is just one rational prime p such that π divides p.

Proof

Let $\pi = a + b\omega$. We have

$$N(\pi) = N(a + b\omega) = (a + b\omega) \cdot (a + b\omega^2) = \pi \cdot (a + b\omega^2) = a^2 - ab + ab.$$

So π divides the positive integer $N(\pi)$. Let p denote the least such positive integer divisible by π. We claim that p is a rational prime. For otherwise there would be a positive integer less than p which is divisible by π. If π divides two distinct rational primes then it divides their GCD which is 1 and so π would be a unit and not a prime in $\mathbb{Z}[\omega]$.

Lemma 3

Any rational prime p is either a prime π in $\mathbb{Z}[\omega]$ or a product $\pi_1 \cdot \pi_2$ (distinct or not) of two primes π_1 and π_2 in $\mathbb{Z}[\omega]$.

Proof

Let p be a rational prime. If p is not a prime in $\mathbb{Z}[\omega]$ then it is divisible by some prime $\pi = a + b\omega$ of $\mathbb{Z}[\omega]$. Let $\pi \in \mathbb{Z}[\omega]$ be a prime such that π divides p. Let $p = \pi \cdot \alpha$ for some $\alpha \in \mathbb{Z}[\omega]$. Then $p^2 = N(\pi) \cdot N(\alpha)$. If $N(\alpha) = 1$ then α is a unit and p is an associate of the prime π. If $N(\alpha) = p$ then α is a prime in $\mathbb{Z}[\omega]$ and p is the product of two primes in $\mathbb{Z}[\omega]$, namely $p = N(\pi) = (a + b\omega) \cdot (a + b\omega^2)$.

Lemma 4

The rational primes which are primes in $\mathbb{Z}[\omega]$ together with the prime factors in $\mathbb{Z}[\omega]$ of other rational primes constitutes all the primes in $\mathbb{Z}[\omega]$.

Proof

If π is any prime in $\mathbb{Z}[\omega]$ then there is a unique rational prime p such that π divides p. So either π is a rational prime or is a factor of some rational prime. Hence, the totality of those rational primes which happen to be prime in $\mathbb{Z}[\omega]$ together with those primes in $\mathbb{Z}[\omega]$ which appear as factors of any rational primes constitute all the primes in $\mathbb{Z}[\omega]$. Now all the primes in $\mathbb{Z}[\omega]$ are easily described. First, we note that since there are six units in $\mathbb{Z}[\omega]$, namely, $\pm 1, \pm\omega, \pm\omega^2$, each prime π in $\mathbb{Z}[\omega]$ has six associates (including itself). A (positive) rational prime $p \in \mathbb{Z}$, $p \equiv 2 \pmod 3$ is along with its associates is a prime in $\mathbb{Z}[\omega]$. The rational prime 3 is an associate of square of prime in $\mathbb{Z}[\omega]$.

$$3 = (-1)(\sqrt{-3} \cdot \sqrt{-3}) = (1-\omega)(1-\omega^2) = -\omega^2(1-\omega)^2 = -\omega(1-\omega^2)^2$$

So $\pi = 1 - \omega$ is a prime and its six associates are $\pm(1-\omega)$, $\pm\omega(1-\omega)$, and $\pm\omega^2(1-\omega)$. A rational prime p, $p \equiv 1 \pmod 3$ splits up as a product of two distinct nonassociate primes in $\mathbb{Z}[\omega]$. We have seen that a prime $p \in \mathbb{Z}$, $p \equiv 1 \pmod 3$ can be expressed as $a^2 + 3b^2$ for some $a, b \in \mathbb{Z}$. So for a rational prime $p \equiv 1 \pmod 3$, we have

$$p = (a^2 + 3b^2) = (a + b\sqrt{-3})(a - b\sqrt{-3}).$$

By Lemma 1, both $a + b\sqrt{-3}$ and $a - b\sqrt{-3}$ are primes in $\mathbb{Z}[\omega]$ and are nonassociates. So the primes in $\mathbb{Z}[\omega]$ are as follows:

- The rational primes $p \equiv 2 \pmod 3$.
- $\sqrt{-3} \sim 1 - \omega$ is a prime in $\mathbb{Z}[\omega]$ as $3 = (-1)\sqrt{-3} \cdot \sqrt{-3}$.
- The rational primes $p \equiv 1 \pmod 3$ factors as a product to two nonassociate primes in $\mathbb{Z}[\omega]$. This is so because such primes can be expressed as

$$p = a^2 + 3b^2 = (a + b\sqrt{-3})(a - b\sqrt{-3}).$$

Theorem The Diophantine equation $x^3 + y^3 = z^3$ has no nontrivial solutions.

Proof (Euler)

Suppose that the above Diophantine equation has a solution. Without loss of generality we may assume that $gcd(x, y) = gcd(y, z) = gcd(x, z) = 1$. Evidently, two of the x, y, z must be odd and the third even. We may assume that z is even and x, y odd. Also we may assume that $z > 0$ and $x > |y|$. We prove this theorem by Fermat's Method of Descent. This method consists in showing that given a solution (x, y, z) of the above Diophantine equation we can find another solution (x_1, y_1, z_1) satisfying the same conditions and an additional condition, namely, $z_1 < z$. Since $z > 0$, this process cannot continue indefinitely and therefore our initial assumption must be wrong.

Note that

$$(A + \sqrt{-3}B) \cdot (C + \sqrt{-3}D) = (AC - 3BD) + (AD + BC)\sqrt{-3}$$

$$(A + \sqrt{-3}B) \cdot (C - \sqrt{-3}D) = (AC + 3BD) - (AD - BC)\sqrt{-3}.$$

Hence

$$(A^2 + 3B^2)(C^2 + 3D^2) = (AC - 3BD)^2 + 3(AD + BC)^2 = (AC + 3BD)^2 + 3(AD - BC)^2.$$

This is a special case of the following:

$$(A^2 + nB^2)(C^2 + nD^2) = (AC + nBD)^2 + n(AD - BC)^2 = (AC - nBD)^2 + n(AD + BC)^2.$$

The last identity has been called as **Brahmagupta Identity** by Weil (1984, p. 204).

Main Lemma

Let $a, b \in \mathbb{Z}$ be such that $gcd(a, b) = 1$ and $a^2 + 3b^2$ is a perfect cube in $\mathbb{Z}$. Then there exist integers $u, v \in \mathbb{Z}$ such that

$$a + b\sqrt{-3} = (u + v\sqrt{-3})^3, \quad \text{and} \quad a - b\sqrt{-3} = (u - v\sqrt{-3})^3.$$

and therefore

$$a = u(u^2 - 9v^2) = u(u + 3v)(u - 3v), \qquad b = 3v(u^2 - v^2) = 3v(u + v)(u - v)$$

and

$$a^2 + 3b^2 = (u^2 + 3v^2)^3.$$

Proof Since $gcd(a, b) = 1$ both a, b are not even. Both cannot be odd either, for if it were so, we would have $a^2 + 3b^2 \equiv 4 \pmod 8$ and this is not possible since $a^2 + 3b^2$ is a cube. Therefore, a and b have different parity and so $a^2 + 3b^2$ is odd. We write

$$a^2 + 3b^2 = (a + b\sqrt{-3}) \cdot (a - b\sqrt{-3}).$$

We claim that $(a+b\sqrt{-3})$, and $(a-b\sqrt{-3})$ are relatively prime in $\mathbb{Z}[\omega]$. Suppose that $\alpha \in \mathbb{Z}[\omega]$ be such that it divides $(a+\sqrt{-3}b)$ and $(a-\sqrt{-3}b)$. Therefore, it divides their sum and difference. So we have $\alpha|2a$, and $\alpha|2\sqrt{-3}$. Since a^2+3b^2 is odd and is divisible by $N(\alpha)$ it follows that $N(\alpha)$ is odd. We have

$$N(\alpha)|(a^2+3b^2), \quad N(\alpha)|4a^2 \quad \Longrightarrow \quad N(\alpha)|a^2, \text{ and } N(\alpha)|3b^2.$$

Since $gcd(a,b)=1$, this implies that $N(\alpha)=1$ or 3. Suppose $N(\alpha)=3$. Then $3|a$ and $3|(a^2+3b^2)$. Since a^2+3b^2 is a cube, it follows that $3^3|(a^2+3b^2)$. Since $3|a \Rightarrow 3^2|a^2$, we have

$$3^2|a^2, \quad 3^2|(a^2+3b^2) \Rightarrow 3^2|3b^2 \Rightarrow 3|b.$$

But this contradicts the assumption that $gcd(a,b)=1$. Therefore $N(\alpha)=1$. Thus, α is a unit, and $(a+b\sqrt{-3})$ and $(a-b\sqrt{-3})$ are relatively prime in $\mathbb{Z}[\omega]$. Let $a^2+3b^2=M^3$. Since a and b have different parity, M is odd. Evidently, M is not divisible by 3, as otherwise 3 would divide both a and b and this will contradict that $gcd(a,b)=1$. We claim that if p is a prime divisor of M then $p \equiv 1 \pmod 3$. This is true because

$$a^2+3b^2 \equiv 0 \pmod p \Rightarrow \left(\frac{-3}{p}\right) = 1 \Rightarrow p \equiv 1 \pmod 3.$$

So when we factor M as a product of primes in $\mathbb{Z}$, each of the prime divisor p appearing in the factorization is such that $p \equiv 1 \pmod 3$. Let

$$M = p_1 \cdot p_2 \cdot p_3 \cdots p_k, \qquad p_i \equiv 1 \pmod 3, 1 \le i \le k.$$

The primes $p_i's$ need not be distinct. We have

$$a^2+3b^2 = M^3 = p_1^3 \cdot p_2^3 \cdot p_3^3 \cdots p_k^3$$

and

$$p_i = x_i^2+3y_i^2 = (x_i+y_i\sqrt{-3})\cdot(x_i-y_i\sqrt{-3}), \qquad 1 \le i \le k$$

where $(x_i+y_i\sqrt{-3})$ and $(x_i-y_i\sqrt{-3})$ are distinct (nonassociate) primes in $\mathbb{Z}[\omega]$. So we can write

$$a^2+3b^2 = M^3 = \prod_{i=1}^{k} p_i^3 = \prod_{i=1}^{k}(x_i+y_i\sqrt{-3})^3(x_i-y_i\sqrt{-3})^3.$$

That is

$$(a+b\sqrt{-3})(a-b\sqrt{-3}) = \prod_{i=1}^{k} p_i^3 = \prod_{i=1}^{k}(x_i+y_i\sqrt{-3})^3(x_i-y_i\sqrt{-3})^3.$$

Since $x_i \pm y_i\sqrt{-3}$, $1 \le i \le k$ is a prime in $\mathbb{Z}[\omega]$ and $gcd(a+b\sqrt{-3}, a-b\sqrt{-3}) = 1$, we conclude that either $x_i + y_i\sqrt{-3}$ divides $a+b\sqrt{-3}$ or it divides its conjugate $a - b\sqrt{-3}$ for $1 \le i \le k$. Evidently,

$$(x_i + y_i\sqrt{-3}) \mid (a \pm b\sqrt{-3}) \Rightarrow (x_i - y_i\sqrt{-3}) \mid (a \mp b\sqrt{-3}).$$

So $(a+b\sqrt{-3})$ is a product of primes $x_i \pm y_i\sqrt{-3}$ each of which appears with exponent 3. Since a product of two numbers of the type $A+B\sqrt{-3}$ can be put in the same form it follows that

$$a+b\sqrt{-3} = (u+v\sqrt{-3})^3, \text{ and } a-b\sqrt{-3} = (u-v\sqrt{-3})^3.$$

A word of caution here. Properly speaking we should write

$$a+b\sqrt{-3} = \lambda(u+v\sqrt{-3})^3, \text{ and } a-b\sqrt{-3} = \overline{\lambda}(u-v\sqrt{-3})^3$$

where λ is a unit. Since -1 is a cube we may assume that $\lambda = \omega$ or $\lambda = \omega^2$. Suppose $\lambda = \omega$. Then we can write

$$\omega^2(a+b\sqrt{-3}) = (u+v\sqrt{-3})^3, \text{ and } \omega(a-b\sqrt{-3}) = (u-v\sqrt{-3})^3.$$

Now

$$\omega^2(a+b\sqrt{-3}) = \left(\frac{-1-\sqrt{-3}}{2}\right)(a+b\sqrt{-3}) = \frac{-a+3b}{2} - \frac{a+b}{2}\sqrt{-3}.$$

Since a and b have different parity $\frac{-a+3b}{2}$, and $\frac{a+b}{2}$ are not integers and so the above equality cannot hold. The case when $\lambda = \omega^2$ is dealt likewise.

Multiplying these two numbers we get

$$a^2 + 3b^2 = (u^2+3v^2)^3.$$

Also,

$$a+b\sqrt{-3} = (u+v\sqrt{-3})^3 = (u^3 - 9uv^2) + (3u^2v - 3v^3)\sqrt{-3},$$

thus

$$a = u^3 - 9uv^2 = u(u^2 - 9v^2), \quad b = 3u^2v - 3v^3 = 3v(u^2 - v^2).$$

Proof of the main theorem

Let

$$x^3 + y^3 = z^3$$

where $gcd(x, y) = gcd(y, z) = gcd(z, x) = 1$. We also assume that z is positive and even, x and y are odd and $x > |y|$. So both $x + y$ and $x - y$ are even. We let $x + y = 2p$ and $x - y = 2q$ and so $x = p + q$ and $y = p - q$. Since $gcd(x, y) = 1$, we get $gcd(p, q) = 1$. As both x and y are odd, p and q have different parity. Also both p and q are positive. We write

$$x^3 + y^3 = (p + q)^3 + (p - q)^3 = 2p(p^2 + 3q^2) = z^3.$$

Since z is even $z^3 \equiv 0 \pmod 8$. As p and q have different parity, $p^2 + 3q^2$ is odd. Therefore, 8 divides $2p$ and so $4|p$. Since p and q have different parity, q is odd. Dividing by 8, we can write

$$\frac{p}{4}\left(p^2 + 3q^2\right) = \left(\frac{z}{2}\right)^3.$$

Now $gcd\left(p, p^2 + 3q^2\right) = 1$ or 3 because $gcd(p, q) = 1$. Now two cases arise.

Case I $gcd\left(\frac{p}{4}, p^2 + 3q^2\right) = 1$ (so in particular z is not divisible by 3)

Since $\frac{p}{4}\left(p^2 + 3q^2\right) = \left(\frac{z}{2}\right)^3$ both $\frac{p}{4}$ and $(p^2 + 3q^2)$ are perfect cubes. Let

$$\frac{p}{4} = r^3, \qquad p^2 + 3q^2 = s^3.$$

Since $p^2 + 3q^2$ is a cube, by our earlier discussion (Main Lemma), we infer that there exist integers u and v such that

$$p^2 + 3q^2 = (u^2 + 3v^2)^3, \qquad p = u(u + 3v)(u - 3v), \qquad q = 3v(u^2 - v^2).$$

Note that $gcd(u, v) = 1$ because $gcd(p, q) = 1$. Since q is odd, u and v have different parity and v is odd and u even. We write

$$r^3 = \frac{p}{4} = \frac{u}{4}(u + 3v)(u - 3v).$$

Therefore $4|u$. We claim that $gcd(u + 3v, u - 3v) = 1$. Suppose that d divides both $(u + 3v)$ and $(u - 3v)$. Then d must be odd, $d|2u$ and $d|6v$ and so $d|u$ and $d|3v$. Since $gcd(u, v) = 1$ we have $d = 1$ or $d = 3$. If $d = 3$ then $3|u$ which implies that $3|p$, since

$$\frac{p}{4}\left(p^2 + 3q^2\right) = \left(\frac{z}{2}\right)^3$$

it follows that $3|z$, which is not true. So $d = 1$ and we have

$$r^3 = \frac{u}{4}(u + 3v)(u - 3v), \quad gcd\left(\frac{u}{4}, u + 3v\right) = gcd(\frac{u}{4}, u - 3v) = gcd(u + 3v, u - 3v) = 1.$$

Therefore, $\frac{u}{4}$, $u + 3v$, and $u - 3v$ are all cubes.

Let

$$\frac{u}{4} = e^3, \qquad u + 3v = f^3, \qquad u - 3v = g^3$$

Then

$$f^3 + g^3 = 2u = 8e^3 = (2e)^3, \quad \text{or} \quad x_1^3 + y_1^3 = z_1^3$$

where $x_1 = f, y_1 = g$, and $z_1 = 2e$. Since $p^3 < 2p(p^2 + 3q^2) = z^3$, we have $p < z$. Finally

$$|z_1| = 2|e| \leq 2|e^3| = \left|\frac{u}{2}\right| < |u| \leq p < z.$$

This completes the proof in case I.

Case II $gcd\left(\frac{p}{4}, p^2 + 3q^2\right) = 3$ (so z is divisible by 3). Since z is even, it is divisible by 6. Since $2p(p^2 + 3q^2) = z^3$, $3|z \Rightarrow 3|p$. We write

$$\frac{p}{4}(p^2 + 3q^2) = \left(\frac{z}{2}\right)^3 \quad \text{as} \quad \frac{p}{4}\left(q^2 + \frac{p^2}{3}\right)3 = \left(\frac{z}{2}\right)^3$$

or

$$\frac{p}{4}\left(q^2 + 3\left(\frac{p}{3}\right)^2\right) = 9\left(\frac{z}{6}\right)^3.$$

As $3|p$ and $3 \nmid q$, therefore 3 does not divide $\left(q^2 + 3(\frac{p}{3})^2\right)$. This implies that $9|\frac{p}{4}$ that is $36|p$. We rewrite the last equality as

$$\frac{p}{36}\left(q^2 + 3\left(\frac{p}{3}\right)^2\right) = \left(\frac{z}{6}\right)^3.$$

Since $gcd(p, q) = 1$, it follows that

$$gcd\left(\frac{p}{36}, q^2 + 3(\frac{p}{3})^2\right) = 1.$$

Therefore, both $\frac{p}{36}$ and $q^2 + 3\left(\frac{p}{3}\right)^2$ are perfect cubes. Let

$$\frac{p}{36} = r^3, \quad \text{and} \quad q^2 + 3\left(\frac{p}{3}\right)^2 = s^3,$$

for some $r, s \in \mathbb{Z}$. Therefore, by the main lemma, there exist integers u and v such that

$$q = u(u^2 - 9v^2), \quad \text{and} \quad \frac{p}{3} = 3v(u^2 - v^2).$$

We note that $gcd(p, q) = 1$ implies $(u, v) = 1$. Hence $(v, u + v) = 1$. Then

$$4r^3 = \frac{p}{9} = \frac{1}{3} \cdot \frac{p}{3} = \frac{1}{3} \cdot 3v(u^2 - v^2) = v(u^2 - v^2) = v(u - v)(u + v)$$

or $-4r^3 = v(v+u)(v-u)$. Since q is odd, u and v have different parity. Since u is odd v is even. We write

$$-8r^3 = 2v(v+u)(v-u).$$

Now since

$$gcd(2v, v+u) = gcd(2v, v-u) = gcd(v+u, v-u) = 1$$

each of $2v$, $v+u$, and $v-u$ must be perfect cubes. Therefore there exist integers e, f, and g such that

$$2v = e^3, \qquad v+u = f^3, \qquad v-u = g^3$$

and so $f^3 + g^3 = 2v = e^3$ or $x_1^3 + y_1^3 = z_1^3$, $x_1 = f$, $y_1 = g$, $z_1 = e$, and z_1 is even (since $e^3 = 2v$). Also

$$|z_1| = |e| < \left|e^3\right| = |2v| \leq |2v(v-u)(v+u)| = 8\left|r^3\right| = 8\frac{p}{36} < p < z.$$

This completes the proof in case II and therefore the proof of the main theorem.

CHAPTER 10

Computational Number Theory

1. Primality Testing

In the first section of this chapter, we discuss two methods for testing primality of a number. They are

1. Solovay–Strassen primality test
2. Rabin–Miller primality test

Both of these tests are probabilistic in the sense that when these declare a number to be a prime number they do so with a probability of an error which can be made as small as we wish by increasing the number of trials in the algorithm. If the test does indeed declare the number to be composite then the number is certainly composite. The algorithms that we shall discuss are primarily meant for implementation on a computer.

An immediate consequence of the uniqueness of prime factorization of a number is that a composite number n must be divisible by a prime number $p \leq \sqrt{n}$. In other words, if a number n is not divisible by any prime $\leq \sqrt{n}$ then n must be a prime number. Since this primality test involves checking the divisibility of n by prime numbers $\leq \sqrt{n}$, it is not very practical.

Composite numbers n which satisfy the condition $a^{n-1} \equiv 1 \pmod{n}$ for all a such that $gcd(a, n) = 1$ are much talked about in computational number theory literature and are called *Carmichael Numbers* in honor of the mathematician who first noted the existence of such numbers. It can be easily shown that a Carmichael number must have three or more distinct prime divisors and must be an odd number. Let $n > 1$ be an odd number and let $n = p_1^{a_1} \cdot p_2^{a_2} \cdots p_k^{a_k}$, where $p_1 < p_2 < \cdots < p_k$ are prime numbers. Let

$$\lambda = lcm\left\{\phi(p_i^{a_i}) : 1 \leq i \leq k\right\}.$$

Then n is a *Carmichael Number* if and only if λ divides $n - 1$. In particular, a Carmichael number must be square free. It has been proved by Alford et al. (1994) that there are infinitely many Carmichael numbers. The Carmichael numbers less than one million are

$$561, 1105, 1729, 2465, 2821, 6601, 8911, 10585,$$
$$15841, 29341, 41041, 46657, 52633, 62745, 63973, 75361.$$

Let n be an odd composite number. We shall say that n satisfies the Fermat's test relative to base b if

1. $b > 1$, $\gcd(b, n) = 1$.
2. $b^{n-1} \equiv 1 \pmod{n}$.

If a composite n satisfies Fermat's test relative to base b then we shall say that n is a pseudoprime with respect to base b. Evidently if n is an (odd) prime number then n is a pseudoprime to any base $b > 1$ such that $gcd(b, n) = 1$. If an odd composite number n fails to satisfy Fermat's test for a single admissible b (i.e., if n is not a Carmichael number) then it will fail the Fermat's test for at least half of the numbers in the interval $[2, n-1]$. The above observation suggests the following probabilistic procedure (which is successful only for non-Carmichael numbers) of testing the primality of a number n. If a random choice of (say) 20 $b's$ in the interval $[2, n-1]$ fails to satisfy the Fermat's test then we can declare that n is a prime number—the probability of error being less than $\frac{1}{2^{20}}$. This primality test is not very helpful as its rather annoying inapplicability for Carmichael numbers. The two tests that we shall now describe have far fewer such exceptions.

1.1. Solovay–Strassen primality test. To describe this test, we first recall Euler's Criterion concerning the Legendre symbol, namely, if p is an odd prime and a any integer then

$$\left(\frac{a}{p}\right) \equiv a^{\frac{p-1}{2}} \pmod{p}.$$

We also recall that the notion of Legendre symbol can be generalized by allowing p to be any odd natural number greater than one and the extended symbol is called *Jacobi symbol.* That is, if $n > 1$ is an odd natural number, $n = p_1 \cdot p_2 \cdots p_k$, a any integer then

$$\left(\frac{a}{n}\right)_J = \left(\frac{a}{p_1}\right) \cdot \left(\frac{a}{p_2}\right) \cdots \left(\frac{a}{p_k}\right)$$

where the symbol $\left(\frac{a}{n}\right)_J$ on the left denotes the Jacobi symbol and the symbols on the right are the Legendre symbol. We ask the question: Given an odd integer n, for what numbers

a the generalization of Euler's criterion is valid? That is for what $a's$

$$\left(\frac{a}{n}\right) \equiv a^{\frac{n-1}{2}} \pmod{n}.$$

For a given odd composite number n, the above relation (*Euler's Criterion*—as we shall continue to call it) is not true for all $a's$—unlike the case of Carmichael numbers and Fermat's test.

Definition: Euler pseudoprimes An odd composite number n is said to be an Euler-pseudo prime to base b if

$$\left(\frac{b}{n}\right)_J \equiv b^{\frac{n-1}{2}} \pmod{n}.$$

Unlike Carmichael numbers, there are no odd composite numbers n which are Euler-pseudoprimes with respect to every possible base. The following theorem shows that an odd composite number n fails to satisfy Euler's criterion for at least half of the numbers in the interval $[1, n-1]$.

Theorem Let n be an odd composite number. Then at least for half of the integers b in the interval $[1, n-1]$ such that $gcd(b, n) = 1$, we have

$$\left(\frac{b}{n}\right)_J \not\equiv b^{\frac{n-1}{2}} \pmod{n}.$$

Proof We shall show that there exists at least one b, $1 < b \leq n-1$, $gcd(b, n) = 1$, and

$$\left(\frac{b}{n}\right)_J \not\equiv b^{\frac{n-1}{2}} \pmod{n}.$$

First suppose that n is square-free. So n is a product of distinct primes. Let $n = p_1 \cdot p_2 \cdots p_k$, $k > 1$, and $p_1 < p_2 < \cdots < p_k$ are prime numbers. Let a be quadratic nonresidue $\pmod{p_1}$. With the help of the Chinese Remainder Theorem, we can find an integer b such that $b \equiv a \pmod{p_1}$ and $b \equiv 1 \pmod{p_i}$, $2 \leq i \leq k$. So

$$\left(\frac{b}{n}\right)_J = \left(\frac{b}{p_1}\right) \cdot \left(\frac{b}{p_2}\right) \cdots \left(\frac{b}{p_k}\right) = -1.$$

We claim that b does not satisfy Euler's criterion. Suppose that

$$\left(\frac{b}{n}\right)_J \equiv b^{\frac{n-1}{2}} \pmod{n}.$$

Then $b^{\frac{n-1}{2}} \equiv -1 \pmod{n}$. In particular $b^{\frac{n-1}{2}} \equiv -1 \pmod{p_i}$, $1 \leq i \leq k$. This contradicts the fact that $b \equiv 1 \pmod{p_i}$, $2 \leq i \leq k$. Hence b does not satisfy the Euler's criterion.

Now suppose that n is not square-free. So there exists a prime p such that p^2 divides n. Suppose that p^r (where $r \geq 2$) is the highest power of the prime p dividing n. Since $\phi(p^r) = p^{r-1}(p-1)$ and the group $\mathbb{Z}_{p^r}^*$ is cyclic (recall that n is odd and so is p) we can find an element a in $\mathbb{Z}_{p^r}^*$ which has order p. Again by using the Chinese Remainder Theorem, we can find b such that $b \equiv a \pmod{p^r}$ and $b \equiv 1 \pmod{\frac{n}{p^r}}$. We claim that this b fails to satisfy Euler's criterion. Suppose $\left(\frac{b}{n}\right)_J = b^{\frac{n-1}{2}} \pmod{n}$. Since $\left(\frac{b}{n}\right) = \pm 1$, we have $b^{n-1} \equiv 1 \pmod{n}$. As an element of $\mathbb{Z}_{p^r}^*$, a (and therefore b) has order p, $n-1$ must be divisible by p. This is not possible. Hence

$$\left(\frac{b}{n}\right) \not\equiv b^{\frac{n-1}{2}} \pmod{n}.$$

So we have shown that there is at least one integer b in the interval $[1, n-1]$ such that $gcd(b,n) = 1$ and $\left(\frac{b}{n}\right)_J \not\equiv b^{\frac{n-1}{2}} \pmod{n}$—that is b does not satisfy Euler's criterion for the odd composite integer n. Now suppose that b and b' are two integers in the interval $[1, n-1]$ such that $gcd(b,n) = gcd(b',n) = 1$ and are such that b' satisfies Euler's criterion but b does not satisfy Euler's criterion. We assert that bb' does not satisfy Euler's criterion. For we have

$$\left(\frac{b'}{n}\right) \equiv b'^{\frac{n-1}{2}} \pmod{n}$$

$$\left(\frac{b}{n}\right) \not\equiv b^{\frac{n-1}{2}} \pmod{n}$$

and $gcd(b,n) = gcd(b',n) = 1$. This immediately implies that

$$\left(\frac{bb'}{n}\right) \not\equiv (bb')^{\frac{n-1}{2}} \pmod{n}.$$

Trivially $b = 1$ satisfies Euler's criterion for n. Suppose $\{b_1, b_2, \cdots, b_t\}$, where $gcd(b_i, n) = 1$, $1 \leq i \leq t$, is the set of *all* integers in the interval $[1, n-1]$ which satisfy Euler's criterion for n. What we have shown above amounts to saying that $\{b_1, b_2, \cdots, b_t\}$ is a proper subset of the set of integers in $[1, n-1]$. Let b be an integer, $1 \leq b \leq n-1$, $gcd(b,n) = 1$ and is such that it *does not* satisfy Euler's criterion for n. Therefore, none of the integers bb_i, $1 \leq i \leq t$ satisfies Euler's criterion for n. Since the numbers bb_i, $1 \leq i \leq t$ are distinct modulo n, it follows that there are at least as many numbers b in the interval $[1, n-1]$ which fail to satisfy Euler's criterion for n as there are which do satisfy Euler's criterion. This completes the proof of the theorem.

1.2. Solovay–Strassen primality test. Suppose that we are given a (fairly large) odd number n. We randomly pick an integer b in the interval $[2, n-2]$. We check whether $gcd(b, n) = 1$. If $gcd(b, n) \neq 1$, then n is composite and we are even able to find a factor of n, namely $gcd(b, n)$. So in this case, the procedure stops here declaring n to be composite. If $gcd(b, n) = 1$ then we test whether b satisfies Euler's criterion for n, namely, whether $\left(\frac{b}{n}\right) \equiv b^{\frac{n-1}{2}} \pmod{n}$. If b does not satisfy Euler's criterion then again n is declared composite and the procedure terminates. If b does satisfy the Euler's criterion we repeat the process with another random b in the interval $[2, n-2]$. If after 20 (say) trials we fail to find some b which does not satisfy Euler's criterion then n is declared to be a prime with the probability $1 - \frac{1}{2^{20}}$. This statement should be interpreted to mean that this algorithm could give a wrong result (namely, declare a composite number to be a prime) not more than once in a million ($\approx 2^{20}$). If for an odd number n, we perform 50 or 100 iterations of this algorithm and still fail to find a b which does not satisfy Euler's criterion then we can be virtually certain that this odd number n is a prime number—as the probability of error would be (in case of 100 trials) less than $\frac{1}{2^{100}}$.

Note: Although in the previous theorem we have proved that for an odd composite integer n at least half of the numbers b in the interval $[2, n-1]$ fail to satisfy Euler's criterion in reality the proportions of such $b's$ (which are sometimes referred to as *witness to the compositeness of* n in the literature) is significantly large. For example, in case of the composite number $n = 77$ or 91 more than 75% of the integers in the interval $[1, n-1]$ fail to satisfy Euler's criterion.

1.3. Rabin–Miller primality test. We shall first describe the main idea underlying this test. If p is a prime number then integers modulo p form a field. We know that over a field a polynomial of degree k has at most k zeros. In particular, the congruence $x^2 \equiv 1 \pmod{p}$ has precisely two solutions, namely 1 and $p-1$, which are distinct if $p > 2$. If for some integer n, the congruence $x^2 \equiv 1 \pmod{n}$ has more than two solutions then n cannot be a prime number. This observation is the basic idea underlying the Rabin–Miller primality test.

Let $n > 2$ be a large odd number. We write $n - 1 = 2^t m$ where m is odd. Let $b > 1$ be any integer such that $gcd(b, n) = 1$. Let $a \equiv b^m \pmod{n}$. Consider the following sequence of numbers (modulo n):

$$a, \quad a^2, \quad a^{2^2}, \quad a^{2^3}, \quad \cdots\cdots, a^{2^t} \qquad \textbf{(A)}$$

Since $a^{2^t} = b^{2^t m} = b^{n-1}$, if n is a prime number then by Fermat's Little Theorem $a^{2^t} \equiv 1 \pmod{n}$. So if n is a prime number then the last number in the sequence **(A)** above is congruent to 1 modulo n. There are four possible types in regard to the sequence of numbers indicated above. These possible types are

1. All numbers in the sequence are congruent to 1 modulo n.
2. One of the numbers in the sequence (before the last one) is congruent to -1 modulo n and (therefore) all the subsequent numbers in the sequence are congruent to 1 mod n.
3. One of the numbers in the sequence (after) the first one is congruent to 1 mod n and the number immediately preceding it is *not* congruent to $\pm 1 \pmod{n}$.
4. None of the numbers in the sequence is congruent to $1 \pmod{n}$.

If n is a prime number then the sequence **(A)** will be of type (1) or (2). In other words, if for some odd number n the sequence **(A)** is of type (3) or (4) then n must be a composite number; however, for odd composite numbers n the sequence **(A)** may turn out to be of any of the four types for different choices of b.

Definition

An odd number $n > 1$ is said to be a *strong pseudo prime* to base b if $gcd(b, n) = 1$ and the sequence **(A)** for the number n (corresponding to base b) is of type (1) or type (2).

It can be shown that an odd composite number n is a strong pseudo prime with respect to base b for at most 25% of the $b's$ in the interval $(1, n-1)$. So roughly speaking for an odd composite number n the probability that a randomly chosen b in the interval $(1, n-1)$ gives rise to the sequence **(A)** of type (3) or (4) is greater than 0.75. We shall refer to such $b's$ in the interval $(1, n-1)$ as witness to the compositeness of the integer n. The above description suggests the following method of testing primality of a number. This procedure is known as Rabin–Miller primality test.

Let n be a (large) odd number. Let $n - 1 = 2^t m$, where m is odd.

Step 1: Randomly pick a number b in the interval $(1, n-1)$. Check that $(b, n) = 1$. If $gcd(b, n) \neq 1$ then n is composite and the algorithm terminates.

Step 2: Let $a \equiv b^m \pmod{n}$. Compute the following sequence of numbers $\pmod{n}$:

$$a, \; a^2, \; a^{2^2}, \; a^{2^3}, \; \cdots\cdots, a^{2^t}.$$

If the above sequence of numbers is of type (3) or (4) then n is a composite and algorithm terminates.

If the above sequence of numbers is of type (1) or (2) then the above two steps are repeated with a different choice of b in the interval $(1, n-1)$. If 20 (say) random choices of $b's$ in the interval $(1, n-1)$ fail to produce a sequence of type (3) or (4) then the number n is declared to be a prime number—with a probability of error less than $\frac{1}{4^{20}}$ (i.e., less than one trillionth!).

The following table gives the value of the smallest odd composite number n which fails the Rabin–Miller test [i.e., for which the sequence **(A)** is of type (3) or (4)] for various values of small prime base b. In this table, x represents the smallest odd composite number n such that n is a strong pseudo prime to the bases less than b but not to base b.

b	x		
3	2047	=	$23 \cdot 89$
5	1373653	=	$829 \cdot 1657$
7	25326001	=	$2251 \cdot 11251$
11	3215031751	=	$151 \cdot 751 \cdot 28351$
13	2152302898747	=	$6763 \cdot 10627 \cdot 29947$
17	3474749660383	=	$1303 \cdot 16927 \cdot 157543$

Source: Taken from Jaeschke (1993).

So in order to test the primality of an odd number n less than $3,474,749,660,383$ all we have to do is to try bases $2, 3, 5, 7, 11$, and 13 in the Rabin–Miller algorithm. If n passes the test with respect to these bases [i.e., the corresponding sequence is of type (1) or (2)] then n is a prime number. A composite odd number $n < 3,474,749,660,383$ will certainly give rise to the corresponding **(A)** sequence of type (3) or (4) for some base b in $\{2, 3, 5, 7, 11, 13\}$. That is to say that in six or fewer iterations we can conclude the primality or compositeness of all numbers less than 3.47×10^{13}.

2. RSA Cryptography

RSA (after Rivest, Shamir, and Adleman) cryptography is one of the much talked about and probably the most widely used public key cryptography scheme today. The word public key means that the method of encryption is known to all and a given message can be encrypted according to some public-key which can be obtained from sources similar to a telephone directory or an Internet directory. The method of encryption is thus publicly known. However, the method of decryption is based upon a secret key known only to the owner. In the absence of the knowledge of this secret decoding key, one can only employ a brute force method to decode the given encrypted message—a process that may take such a long time (many years) so as to make the message practically useless for a potential adversary. The basic idea behind the RSA cryptography is inherent in the observation that there is no presently known "fast" algorithm for factorization of large integers. If sometime in the future someone comes up with a "fast" algorithm (i.e., an algorithm whose running time is a polynomial function of the size of the input) for factorization of large integers then the RSA cryptographical scheme will lose its practical usefulness. We shall now describe the RSA public-key cryptographic scheme.

Let p and q be two distinct, large odd prime numbers. Let $n = pq$. Choose e such that $gcd(e, \phi(n)) = 1$, where ϕ is Euler's totient function. Let d be such that $de \equiv 1 \pmod{\phi(n)}$. Let x be any integer $0 \leq x < n$. We regard x as the message to be encoded. We compute y, $0 \leq y < n$, $y \equiv x^e \pmod{n}$. Then y constitutes the encrypted message. The pair (e, n) is considered to be the encryption key and it is this pair (together with the block sizes to be described soon) which is in the public domain. The receiver (the owner of the public key) after receiving the message y has only to raise y to the power d modulo n to retrieve the message x. Since $de \equiv 1 \pmod{\phi(n)}$, it is clear that

$$y^d \equiv (x^e)^d = x^{de} \equiv x \pmod{n}.$$

The message to be encrypted is divided into blocks of equal size and each block (or rather its certain numerical equivalent) is encrypted individually. The block sizes (i.e., the encryption block size and decryption block size) are parts of the encryption process and therefore are components of the public key. We shall denote these block sizes by the letters k and ℓ, respectively, and assume that $k < \ell$.

Suppose that we agree to work with the 30-letter alphabet consisting of 26 upper case roman letters, together with the symbols, blank, period, comma, and question mark. The roman letters are assigned values from 0 to 25, blank = 26, period = 27, comma = 28, and question mark = 29.

Suppose we agree to divide the message to be encrypted into block size seven and the encrypted message block size be eight. So $k = 7$ and $\ell = 8$. We shall need two (large) prime numbers p and q such that $n = pq$ satisfies $30^7 < n < 30^8$. We let $p = 636,319$ and $q = 879,817$. So $n = pq = 559,844,273,623$. We note that n satisfy the inequality $30^7 < n < 30^8$. We let $e = 996,173$. We have chosen e to be a prime number greater than p and q. This ensures that $gcd(e, \phi(n)) = 1$. After some computation using Euclidean algorithm we will find that d, the inverse of e modulo $\phi(n)$, is $223,259,788,421$. Let us encrypt the following message:

RSA□CRYPTOGRAPHY□IS□VERY□INTERESTING.□IS□IT□NOT?□

We note that in the message to be encrypted any two intermediate words are separated by (one) blank which we have denoted by the symbol □. Since length of the given message (including intermediate blanks) is 48 characters we pad the given message with one blank on the right so that the total message length is evenly divisible by the input block length k which we have taken to be seven. So the given input message is divided into blocks of size seven which gives us seven blocks in all (just a coincidence). We first compute the numerical equivalent of each block by considering it as a base 30 number. For example, the first block, namely

RSA□CRY

corresponds to the number

$$17 \cdot 30^6 + 18 \cdot 30^5 + 0 \cdot 30^4 + 26 \cdot 30^3 + 2 \cdot 30^2 + 17 \cdot 30 + 24$$

which is equal to $12,831,104,334$. This number when raised to power $e = 996,173$, and reduced modulo $n = 559,844,273,623$, gives us $405,817,953,114$. This number can be expressed in base 30 as

$$18 \cdot 30^7 + 16 \cdot 30^6 + 20 \cdot 30^5 + 9 \cdot 30^4 + 24 \cdot 30^3 + 16 \cdot 30^2 + 23 \cdot 30 + 24.$$

This corresponds to the encrypted block of size eight which is $SQUJYQXY$. The receiver first converts the block of eight characters into its equivalent base 30 number which (of course) turns out to be $405,817,953,114$ and raises it to power $d = 223,953,114$ modulo n. This gives the receiver the number $12,831,104,334$—which as we have seen, is the first (message) block of length seven namely $RSA\Box CRY$ of the original message. The accompanied table gives the encrypted equivalent of all the seven blocks in the above message. We have

$p = 636319$, $q = 879817$, $n = 559844273623$, $e = 996173$, $d = 223259788421$.

Input String:

RSA□CRYPTOGRAPHY□IS□VERY□INTERESTING.□IS□IT□NOT?□

Encrypted String:

SQUJYQXYT,GEXWBUKD,R□NCJHUBPO,ZNIBAHMTN?VHLMQJXYJ,VP□XFT

Decrypted Message:

RSA□CRYPTOGRAPHY□IS□VERY□INTERESTING.□IS□IT□NOT?□

The length of the message is 49 (this includes a blank in the end after the question mark "?"). Input block length = 7. Output block length = 8. The number of blocks is 7.

Block Number	Input Block Contents	Numerical Equivalent	Numerical Encryption	Output Block Contents
1	RSA□CRY	12831104334	405817953114	SQUJYQXY
2	PTOGRAP	11408217315	436091680850	T,GEXWBU
3	HY□IS□V	5707493001	221581883769	KD,R□NCJ
4	ERY□INT	3349249609	167706853963	HUBPO,ZN
5	ERESTIN	3332843353	175695011519	IBAHMTN?
6	G.□IS□I	5051392988	464650460814	VHLMQJXY
7	T□NOT?□	14493725996	217765172869	J,VP□XFT

3. Computing $a^n \pmod{m}$ by Repeated Squaring

To compute $a^n \pmod{m}$, we first express n in binary. Suppose that $n = b_1b_2b_3\cdots b_k$ in binary. So $b_i \in \{0,1\}$ and $b_1 = 1$. We fill in the following table whose second column titled "Binary Digits" consists of the k binary digits, of the number n beginning from left to right, so the entry in the ith row is the ith binary digit of n. The ith entry in the third column is the decimal equivalent of the binary number consisting of the first (beginning from left) i binary digits of n. The ith entry in the last column is the square (mod m) of the previous entry in the same column if the ith binary digit is zero and is the square of the previous entry in the same column times a mod m if the ith entry is one. We shall illustrate this by two examples.

Example 1 Let us calculate $19^{163} \pmod{59}$. So here $a = 19, n = 163$, and $m = 59$. We have $n = 163 = 128+32+2+1 = 10100011$ (in binary). The entries in the last column are reduced modulo 59.

i	Binary Digit (b_i)	$x = b_1b_2\cdots b_i$	$19^x \pmod{59}$
1	1	1	19
2	0	2	$19^2 = 361 \equiv 7$
3	1	5	$19^5 \equiv 7^2 \times 19 \equiv 46$
4	0	10	$19^{10} \equiv 46^2 \equiv 51$
5	0	20	$19^{20} \equiv 51^2 \equiv 5$
6	0	40	$19^{40} \equiv 5^2 \equiv 25$
7	1	81	$19^{81} \equiv 25^2 \times 19 \equiv 16$
8	1	163	$19^{163} \equiv 16^2 \times 19 \equiv 26$

Therefore $7^{163} \equiv 27 \pmod{59}$.

Example 2 Let us calculate $7^{97} \pmod{239}$. So here $a = 7, n = 97$, and $m = 239$. We have $n = 97 = 64+32+1 = 1100001$ (in binary). The entries in the last column are reduced modulo 239.

i	Binary Digit (b_i)	$x = b_1b_2\cdots b_i$	$7^x \pmod{239}$
1	1	1	7
2	1	3	$7^3 \equiv 104$
3	0	6	$7^6 \equiv 104^2 \equiv 61$
4	0	12	$7^{12} \equiv 61^2 \equiv 136$
5	0	24	$7^{24} \equiv 136^2 \equiv 93$
6	0	48	$7^{48} \equiv 93^2 \equiv 45$
7	1	97	$7^{97} \equiv 45^2 \times 7 \equiv 74$

Therefore $7^{97} \equiv 74 \pmod{239}$.

4. Calculating square root mod p ($p > 2$, a prime)

Let $p > 2$ be a prime and $a \in \mathbb{Z}$, $a \neq 1$ be a quadratic residue mod p. So the congruence $x^2 \equiv a \pmod{p}$ has a solution. We will discuss a method to find a solution of such a congruence. Two cases (i) $p \equiv 1 \pmod 4$ and (ii) $p \equiv 3 \pmod 4$ are dealt separately.

1. Case I. $p \equiv 3 \pmod 4$. In this case, the solution is given by $x \equiv a^{\frac{p+1}{4}} \pmod{p}$. This can be easily checked as follows:

$$\left(a^{\frac{p+1}{4}}\right)^2 = a^{\frac{p+1}{2}} = a^{\frac{p-1}{2}} \cdot a \equiv (x^2)^{\frac{p-1}{2}} \cdot a \equiv x^{p-1} \cdot a \equiv a \pmod{p}.$$

2. Case II. $p \equiv 1 \pmod 4$. In this case, we first find a quadratic nonresidue mod p, say h. Suppose that we can find an odd integer (say $2m-1$) and an even integer (say $2n$) such that $a^{2m-1}h^{2n} \equiv 1 \pmod{p}$ then $x \equiv a^m h^n \pmod{p}$ is a solution to the congruence $x^2 \equiv a \pmod{p}$. The following example illustrates finding such an m and n thereby computing the square root of a mod p in this case.

   Consider $x^2 \equiv 2 \pmod{97}$. Evidently this congruence has a solution. As 5 is a quadratic nonresidue mod 97 we let $h = 5$. We calculate the values of m and n with the help of the following table. In this table $a = 2$, $h = 5$, $e_1 = \frac{p-1}{2} = 48$, and $e_2 = p-1 = 96$. The entries in the columns 3, 4, and 5 are reduced modulo 97. Note that the entries in the fifth column can only be 1 or -1 mod p. Also if $a^{e_1} \cdot h^{e_2} \equiv 1 \pmod{p}$ then $E_2 = e_2$, and if $a^{e_1} \cdot h^{e_2} \equiv -1 \pmod{p}$ then $E_2 = e_2 + \frac{p-1}{2}$. In the first row $e_1 = 48$, $e_2 = E_2 = 96$. In the second row onward (excepting in the last row), e_1 has the value half of its values in the previous row, e_2 has value half of the values of E_2 in the previous row.

e_1	e_2	$a^{e_1} = 2^{e_1}$	$h^{e_2} = 5^{e_2}$	$a^{e_1} \cdot h^{e_2} = 2^{e_1} \cdot 5^{e_2}$	E_2
48	96	1	1	1	96
24	48	96	96	1	48
12	24	22	22	-1	72
6	36	64	50	-1	84
3	42	8	12	-1	90
2	45	4	45	$x \equiv 2^2 \cdot 5^{45} \equiv 4 \cdot 45 \equiv 83 \pmod{97}$	

The solution of the above congruence is $x \equiv 2^2 \cdot 5^{45} \equiv 4 \cdot 45 \equiv 83 \pmod{97}$.

5. Shank's Baby-Step Giant-Step Discrete Log Algorithm

Let p be a prime and $a, b \in \mathbb{Z}$ be such that the congruence $a^x \equiv b \pmod{p}$ has a solution. This algorithm provides a method for solving such a congruence.

Let $[\sqrt{p}] = m$ and $x = m \cdot q + r$, where $0 \leq r < m$. Then

$$b = a^x = a^{m \cdot q + r} = a^{m \cdot q} \cdot a^r, \quad \text{and so} \quad b \cdot a^{-r} \equiv a^{m \cdot q} \equiv (a^m)^q \pmod{p}.$$

So if we can find such q and r, then $x = m \cdot q + r$ is determined. We shall illustrate this method by two examples.

Example 1 Let us solve the congruence $7^x \equiv 74 \pmod{239}$. We note that 239 is a prime number and 7 is a primitive root modulo 239, that is the order of 7 modulo 239 is 238. So this congruence has a unique solution. Here we have

$$a = 7, b = 74, p = 239, m = [\sqrt{239}] = 15, \quad \text{and} \quad 7^{15} \equiv 43 \pmod{239}.$$

r	$b \cdot a^{-r} = 74 \cdot 7^{-r} \equiv 74 \cdot (205)^r \pmod{239}$	q	$(a^m)^q \equiv (7^{15})^q \equiv 43^q \pmod{239}$
0	74	0	1
1	113	1	43
2	221	2	176
3	134	3	159
4	224	4	145
5	32	5	21
6	107	6	**186**
7	**186**	7	111
8	129	8	232
9	155	9	177
10	227	10	202
11	169	11	82
12	229	12	180
13	101	13	92
14	151	14	132
15	124	15	179

So we have $q = 6$ and $r = 7$ and this gives $x = 97$, and as can be easily checked $7^{97} \equiv 74 \pmod{239}$.

Example 2 Let us solve the congruence $5^x \equiv 1303 \pmod{2017}$.

We note that 5 is a primitive root modulo 2017. In this case, we let $m = [\sqrt{2017}]+1 = 45$ and $x = 45q + r$, where $0 \le q, r < 45$. Then $5^x = 5^{45q+r} \equiv 1303 \pmod{2017}$. Since $5^{-1} \equiv 807 \pmod{2017}$ and $5^{45} \equiv 45 \pmod{2017}$, we have $1303 \times 807^r \equiv 45^q \pmod{2017}$. We calculate the values of q and r in the following table knowing that $0 \le q, r < 45$.

r	1303×807^r mod(2017)	r	1303×807^r mod(2017)	q	45^q mod(2017)	q	45^q mod(2017)
0	1303	23	451	0	1	23	1307
1	664	24	897	1	45	24	322
2	1343	25	1793	2	8	25	371
3	672	26	762	3	360	26	559
4	1748	27	1766	4	64	27	951
5	753	28	1160	5	863	28	438
6	554	29	232	6	512	29	1557
7	1321	30	1660	7	853	30	1487
8	1071	31	332	8	62	31	354
9	1021	32	1680	9	773	32	1811
10	1011	33	336	10	496	33	815
11	1009	34	874	11	133	34	369
12	1412	35	1385	12	1951	35	469
13	1896	36	277	13	1064	36	935
14	1186	37	1669	14	1489	37	1735
15	1044	38	1544	15	444	38	1429
16	1419	39	1519	16	1827	39	1778
17	**1494**	40	1514	17	1535	40	1347
18	1509	41	1513	18	497	41	105
19	1512	42	706	19	178	42	691
20	1916	43	948	20	1959	43	840
21	1190	44	593	21	1424	44	**1494**
22	238			22	1553		

So $q = 44$ and $r = 17$. Therefore, $x = 45 \times 44 + 17 = 1997$ and it can be checked $5^{1997} \equiv 1303 \pmod{2017}$.

CHAPTER 11

Some Bachet Equation

In this chapter, we discuss the Diophantine equations of the type $y^2 = x^3 + k$. Such Diophantine equations are known as Bachet equations or Mordell equations in number theory. We show that for many values of $k \in \mathbb{Z}$, the Diophantine equation $y^2 = x^3 + k$ has no solutions. In addition to the properties of the Legendre symbol, we assume the following easily proved congruence conditions on the prime divisors of natural numbers in various congruence classes:

1. If x is odd then $x^2 \equiv 1 \pmod 8$.
2. If x is not divisible by 3 then $x^2 \equiv 1 \pmod 3$.
3. A natural number $N \equiv 3 \pmod 4$ has a prime divisor $p \equiv 3 \pmod 4$.
4. A natural number $N \equiv 3 \pmod 8$ either has a prime divisor $p \equiv 3 \pmod 8$ or it has prime divisors q_1 and q_2 such that $q_1 \equiv 5 \pmod 8$ and $q_2 \equiv 7 \pmod 8$.
5. A natural number $N \equiv 5 \pmod 8$ either has a prime divisor $p \equiv 5 \pmod 8$ or it has prime divisors q_1 and q_2 such that $q_1 \equiv 3 \pmod 8$ and $q_2 \equiv 7 \pmod 8$.
6. A natural number $N \equiv 7 \pmod 8$ either has a prime divisor $p \equiv 7 \pmod 8$ or it has a prime divisors q_1 and q_2 such that $q_1 \equiv 3 \pmod 8$ and $q_2 \equiv 5 \pmod 8$.
7. A natural number $N \equiv 5 \pmod{12}$ either has a prime divisor $p \equiv 5 \pmod{12}$ or it has a prime divisors q_1 and q_2 such that $q_1 \equiv 7 \pmod{12}$ and $q_2 \equiv 11 \pmod{12}$.
8. A natural number $N \equiv 7 \pmod{12}$ either has a prime divisor $p \equiv 7 \pmod{12}$ or it has a prime divisors q_1 and q_2 such that $q_1 \equiv 5 \pmod{12}$ and $q_2 \equiv 11 \pmod{12}$.
9. A natural number $N \equiv 11 \pmod{12}$ either has a prime divisor $p \equiv 5 \pmod{11}$ or it has a prime divisors q_1 and q_2 such that $q_1 \equiv 5 \pmod{12}$ and $q_2 \equiv 7 \pmod{12}$.

- **Example 1:** The Diophantine equation $y^2 = x^3 + 23$ has no solutions.

  We recall that the square of a number is congruent to 0 or 1 (mod 4). If x is even then RHS is congruent to 3 (mod 4). This is not possible. If $x \equiv 3 \pmod 4$ then

RHS is congruent to 2 (mod 4). This too is not possible. Now suppose that $x \equiv 1$ (mod 4). Then we write $y^2 + 4 = x^3 + 27 = (x+3)(x^2 - 3x + 9)$. Now

$$x \equiv 1 \pmod 4 \Longrightarrow x^2 - 3x + 9 \equiv 3 \pmod 4.$$

So there exists a prime p such that $p \equiv 3$ (mod 4) and $x^2 - 3x + 9 \equiv 0$ (mod p). But then we have

$$y^2 + 4 \equiv 0 \pmod p \Longrightarrow \left(\frac{-4}{p}\right) = \left(\frac{-1}{p}\right) = 1.$$

So the fact that the given Diophantine equation has a solution would imply that -1 is a quadratic residue mod p. This is not true.

- **Example 2:** The Diophantine equation $y^2 = x^3 + 45$ has no solutions.

If x is even then the RHS is congruent to 5 (mod 8) which is not possible. So x must be odd. If $x \equiv 1$ (mod 4) then RHS is congruent to 2 (mod 4) which too is not possible. So x cannot be even or congruent to 1 (mod 4). Suppose that $x \equiv 3$ (mod 4). There are two cases to be considered.

Case I $x \equiv 3$ (mod 8). In this case, we write

$$y^2 - 72 = x^3 - 27 = (x-3)(x^2 + 3x + 9).$$

Now

$$x \equiv 3 \pmod 8 \Longrightarrow x^2 + 3x + 9 \equiv 3 \pmod 8.$$

So $x^2 + 3x + 9$ must be divisible by a prime p which is congruent to 3 or 5 (mod 8). Then the assumption that the above Diophantine equation has a solution would imply that 2 is a quadratic residue mod p. This is not true.

Case II $x \equiv 7$ (mod 8). In this case, we write

$$y^2 - 18 = x^3 + 27 = (x+3)(x^2 - 3x + 9).$$

$$x \equiv 7 \pmod 8 \Longrightarrow x^2 - 3x + 9 \equiv 5 \pmod 8.$$

So $x^2 - 3x + 9$ is divisible by a prime p which is congruent to 3 or 5 (mod 8). Then the assumption that the above Diophantine equation has a solution would imply that 2 is a quadratic residue mod p. This is not true.

- **Example 3:** The Diophantine equation $y^2 = x^3 - 5$ has no solutions.

  As x even implies that $y^2 = x^3 - 5 \equiv 3 \pmod 4$, so x is not even. So we assume that x is odd. If $x \equiv 3 \pmod 4$ then $y^2 \equiv 2 \pmod 4$ which is not possible. So we assume that $x \equiv 1 \pmod 4$. We write

$$y^2 + 4 = x^3 - 1 = (x-1)(x^2 + x + 1)$$

  Now

$$x \equiv 1 \pmod 4 \Longrightarrow x^2 + x + 1 \equiv 3 \pmod 4$$

  So $x^2 + x + 1$ must have a prime divisor p such that $p \equiv 3 \pmod 4$. This would imply that -1 is a quadratic residue mod p, which is not true.

- **Example 4:** The Diophantine equation $y^2 = x^3 + 7$ has no solution.

  Evidently, x cannot be even. Also $x \not\equiv 3 \pmod 4$, as that would imply $y^2 \equiv 2 \pmod 4$, which is not possible. So we assume that $x \equiv 1 \pmod 4$. We write

$$y^2 + 1 = x^3 + 8 = (x+2) \cdot (x^2 - 2x + 4).$$

  Now $x \equiv 1 \pmod 4$ implies that $x + 2 \equiv 3 \pmod 4$ and so $x + 2$ must have a prime divisor $p \equiv 3 \pmod 4$. This would imply that -1 is a quadratic residue mod p which is not possible.

- **Example 5:** The Diophantine equation $y^2 = x^3 + 6$ has no solution.

  Evidently, x cannot be even and it also cannot be congruent to $1 \pmod 4$. So we assume that $x \equiv 3 \pmod 4$ or $x \equiv 3, 7 \pmod 8$. As $x \equiv 7 \pmod 8$ implies that $y^2 \equiv 5 \pmod 8$ which is not possible, we must have $x \equiv 3 \pmod 8$. Now we write

$$y^2 + 2 = x^3 + 8 = (x+2)(x^2 - 2x + 4)$$

$$x \equiv 3 \pmod 8 \Longrightarrow x^2 - 2x + 4 \equiv 7 \pmod 8.$$

  So $x^2 - 2x + 4$ must have a prime divisor p which is congruent to 5 or 7 (mod 8). In either case we get a contradiction as follows:

  If $p \equiv 5 \pmod 8$ then $y^2 + 2 \equiv 0 \pmod p$ and

$$1 = \left(\frac{-2}{p}\right) = \left(\frac{2}{p}\right) = -1.$$

If $p \equiv 7 \pmod 8$ then $y^2 + 2 \equiv 0 \pmod p$ and

$$1 = \left(\frac{-2}{p}\right) = -1 \cdot \left(\frac{2}{p}\right) = -1.$$

So the given Diophantine equation has no solutions.

- **Example 6:** The Diophantine equation $y^2 = x^3 + 11$ has no solution.

Evidently, x cannot be even and it also cannot be congruent to $3 \pmod 4$. So we assume that $x \equiv 1 \pmod 4$. Now we write $y^2 + 16 = x^3 + 27 = (x+3) \cdot (x^2 - 3x + 9)$. Now

$$x \equiv 1 \pmod 4 \Longrightarrow x^2 - 3x + 9 \equiv 3 \pmod 4$$

so $x^2 - 3x + 9$ must have a prime divisor p which is congruent to 3 mod 4. We have

$$1 = \left(\frac{-16}{p}\right) = \left(\frac{-1}{p}\right) = -1.$$

Therefore, the given Diophantine equation has no solution.

- **Example 7:** The Diophantine equation $y^2 = x^3 + 13$ has no solutions.

If x is even then we would have $x \equiv 5 \pmod 8$ which is not possible. So x must be odd. If $x \equiv 1 \pmod 4$ then we would have $y^2 \equiv 2 \pmod 8$, which too is not possible. So we assume that $x \equiv 3 \pmod 4$. We assert that $x \equiv 0 \pmod 3$. Now

$$x \equiv 1 \pmod 3 \Longrightarrow y^2 \equiv 2 \pmod 3, \quad \text{which is not possible}$$

$$x \equiv 2 \pmod 3 \Longrightarrow y^2 = x^3 + 13 \equiv 3 \pmod 9, \quad \text{which too is not possible.}$$

So we are left to deal with the possibility $x \equiv 0 \pmod 3$. As $x \equiv 3 \pmod 4$ and $x \equiv 0 \pmod 3$ imply that $x \equiv 3 \pmod{12}$. Now $y^2 - 12 = x^3 + 1 = (x+1)(x^2 - x + 1)$.

$$x \equiv 3 \pmod{12} \Longrightarrow x^2 - x + 1 \equiv 7 \pmod{12}.$$

Therefore, $x^2 - x + 1$ must have a prime factor $p \equiv 5, 7 \pmod{12}$. If $x^2 - x + 1$ has a prime factor $p \equiv 5 \pmod{12}$ then

$$1 = \left(\frac{12}{p}\right) = \left(\frac{3}{p}\right) = \left(\frac{p}{3}\right) = \left(\frac{2}{3}\right) = -1.$$

If $p \equiv 7 \pmod{12}$ then

$$1 = \left(\frac{12}{p}\right) = \left(\frac{3}{p}\right) = -1 \cdot \left(\frac{p}{3}\right) = -1 \left(\frac{1}{p}\right) = -1.$$

So in either case we get a contradiction and therefore the given Diophantine equation has no solution.

- **Example 8:** The Diophantine equation $y^2 = x^3 + 14$ has no solutions.

Evidently, x cannot be even or congruent to 1 (mod 4). So we assume that $x \equiv 3 \pmod 4$ or equivalently $x \equiv 3, 7 \pmod 8$. Now $x \equiv 7 \pmod 8$ implies that $y^2 \equiv 5 \pmod 8$ which is not possible. So we assume that $x \equiv 3 \pmod 8$. We write

$$y^2 + 50 = x^3 + 64 = (x + 4) \cdot (x^2 - 4x + 16).$$

As $x \equiv 3 \pmod 8$ implies that $x^2 - 4x + 16 \equiv 5 \pmod 8$. So $x^2 - 4x + 16$ must be divisible by some prime p which is $\equiv 5 \pmod 8$ or $\equiv 7 \pmod 8$. If $p \equiv 5 \pmod 8$ then

$$1 = \left(\frac{-50}{p}\right) = \left(\frac{2}{p}\right) = -1.$$

If $p \equiv 7 \pmod 8$ then

$$1 = \left(\frac{-50}{p}\right) = -1 \cdot \left(\frac{2}{p}\right) = -1.$$

So in either case we get a contradiction.

- **Example 9:** The Diophantine equation $y^2 = x^3 + 34$ has no solution.

Evidently, x cannot be even and it also cannot be congruent to 1 (mod 4). So we assume that $x \equiv 3 \pmod 4$ or $x \equiv 3, 7 \pmod 8$. As $x \equiv 3 \pmod 8$ implies that $y^2 \equiv 5 \pmod 8$ which is not possible, we must have $x \equiv 7 \pmod 8$. Now we write

$$y^2 - 98 = x^3 - 64 = (x - 4)(x^2 + 4x + 16).$$

Now $x \equiv 7 \pmod 8$ implies that $x - 4 \equiv 3 \pmod 8$. Two cases arise

Case I $x - 4$ has a prime divisor $p \equiv 3 \pmod 8$.

Case II $x - 4$ has prime divisors p and q such that $p \equiv 5 \pmod 8$ and a prime divisor q such that $q \equiv 7 \pmod 8$.

So in any case $x - 4$ has a prime divisor p which is congruent to 3 or 5 (mod 8).

$$1 = \left(\frac{98}{p}\right) = \left(\frac{2}{p}\right) = -1.$$

So in either case we get a contradiction.

- **Example 10:** The Diophantine equation $y^2 = x^3 + 39$ has no solution.

Evidently, x cannot be even. Also $x \equiv 3 \pmod 4$ implies that $y^2 \equiv 2 \pmod 4$ which is not possible. So we assume that $x \equiv 1 \pmod 4$. We write

$$y^2 + 961 = x^3 + 1000 = (x + 10) \cdot (x^2 - 10x + 100).$$

Now $x \equiv 1 \pmod 4$ implies that $x^2 - 10x + 100 \equiv 3 \pmod 4$. But then the assumption that the given Diophantine equation has a solution would imply that -1 is a quadratic residue modulo p, which is not true.

- **Example 11:** The Diophantine equation $y^2 = x^3 + 42$ has no solution.

Evidently, x cannot be even. Also $x \not\equiv 1 \pmod 4$, as otherwise $y^2 \equiv 3 \pmod 4$, which is not possible. So we assume that $x \equiv 3 \pmod 4$. Therefore $x \equiv 3$ or $7 \pmod 8$.

$$y^2 - 50 = x^3 - 8 = (x - 2) \cdot (x^2 + 2x + 4).$$

Now $x \equiv 3, 7 \pmod 8$ implies that $N = x^2 + 2x + 4 \equiv 3 \pmod 8$. Therefore, N must be divisible by some prime p which is either $\equiv 3 \pmod 8$ or $x \equiv 5 \pmod 8$. In either case we get a contradiction from

$$1 = \left(\frac{50}{p}\right) = \left(\frac{2}{p}\right) = -1.$$

- **Example 12:** The Diophantine equation $y^2 = x^3 + 47$ has no solution.

Evidently, x cannot be even. Also $x \not\equiv 3 \pmod 4$, as otherwise $y^2 \equiv 2 \pmod 4$, which is not possible. So we assume that $x \equiv 1 \pmod 4$. We write

$$y^2 + 169 = x^3 + 216 = (x + 6) \cdot (x^2 - 6x + 36).$$

Now $x \equiv 1 \pmod 4$ implies that $x + 6 \equiv 3 \pmod 4$. Therefore $x + 6$ must have a prime divisor p which is congruent to $3 \pmod 4$. Again the assumption that the given Diophantine equation has a solution would imply that -1 is a quadratic residue mod p, which is not true.

- **Example 13:** The Diophantine equation $y^2 = x^3 + 53$ has no solution.

Evidently, x cannot be even as this would imply that $y^2 \equiv 5 \pmod 8$. So x must be odd. If $x \equiv 1 \pmod 4$ then $y^2 \equiv 2 \pmod 4$ which is not possible. So we assume that $x \equiv 3 \pmod 4$. We write

$$y^2 + 676 = x^3 + 729 = (x+9) \cdot (x^2 - 9x + 81).$$

Now $x \equiv 3 \pmod 4$ implies that $x^2 - 9x + 81 \equiv 3 \pmod 4$ and we get the usual contradiction.

- **Example 14:** The Diophantine equation $y^2 = x^3 + 46$ has no solution.

Evidently, x cannot be even as this would imply that $y^2 \equiv 6 \pmod 8$. So x must be odd. If $x \equiv 1 \pmod 4$ then $y^2 \equiv 3 \pmod 4$ which is not possible. So we assume that $x \equiv 3 \pmod 4$ and therefore either $x \equiv 3 \pmod 8$ or $x \equiv 7 \pmod 8$. But if $x \equiv 7 \pmod 8$ then $y^2 \equiv 5 \pmod 8$ which is not possible. So we assume that $x \equiv 3 \pmod 8$. We write

$$y^2 + 18 = x^3 + 64 = (x+4) \cdot (x^2 - 4x + 16).$$

$x \equiv 3 \pmod 8$ implies that, $x + 4 \equiv 7 \pmod 8$. Two case arise.

Case I $x + 4$ has a prime divisor p such that $\equiv 7 \pmod 8$.

In this case, we get a contradiction from

$$1 = \left(\frac{-18}{p}\right) = -\left(\frac{2}{p}\right) = -1.$$

Case II $x + 4$ has no prime divisor which is $\equiv 7 \pmod 8$ so it must have prime divisors p and q such that $p \equiv 3 \pmod 8$ and $q \equiv 5 \pmod 8$. In this case, we get a contradiction from

$$1 = \left(\frac{-18}{q}\right) = \left(\frac{2}{q}\right) = -1.$$

- **Example 15:** The Diophantine equation $y^2 = x^3 + 20$ has no solution.

First suppose that x is odd. As $x \equiv 3 \pmod 4$ would imply that $y^2 \equiv 3 \pmod 4$, we may assume that $x \equiv 1 \pmod 4$. We write

$$y^2 + 196 = x^3 + 216 = (x+6) \cdot (x^2 - 6x + 36).$$

As $x + 6 \equiv 3 \pmod 4$ so it must have a prime divisor $p \equiv 3 \pmod 4$ and we get the usual contradiction.

Now suppose that x is even. Then y too must be even. Let $x = 2x_1$ and $y = 2y_1$. So we have $y_1^2 = 2x_1^3 + 5$. If x_1 is even then $y_1^2 \equiv 5 \pmod 8$ and if x_1 is odd then $y_1^2 \equiv 3$ or $7 \pmod 8$. So in either case we get a contradiction.

- **Example 16:** The Diophantine equation $y^2 = x^3 + 21$ has no solution.

Evidently, x cannot be even or $\equiv 1 \pmod 4$. So we assume that $x \equiv 3 \pmod 4$. Evidently $x \not\equiv 0, 2 \pmod 3$. So we may also assume that $x \equiv 1 \pmod 3$. We write

$$y^2 - 48 = x^3 - 27 = (x - 3) \cdot (x^2 + 3x + 9).$$

Let $N = x^2 + 3x + 9$. Now $x \equiv 1 \pmod 3$ and $x \equiv 3 \pmod 4$ imply that $x \equiv 7 \pmod{12}$, which in turn imply that $N \equiv 7 \pmod{12}$. Now two cases arise

Case I N has a prime divisor $p \equiv 7 \pmod{12}$. In this case we have the following contradiction.

$$1 = \left(\frac{48}{p}\right) = \left(\frac{3}{p}\right) = -\left(\frac{p}{3}\right) = -\left(\frac{1}{3}\right) = -1.$$

Case II N has a prime divisor $p \equiv 5 \pmod{12}$ and a prime divisor $q \equiv 11 \pmod{12}$. In this case we have the following contradiction.

$$1 = \left(\frac{48}{p}\right) = \left(\frac{3}{p}\right) = \left(\frac{p}{3}\right) = \left(\frac{2}{3}\right) = -1.$$

- **Example 17:** The Diophantine equation $y^2 = x^3 - 10$ has no solutions.

Evidently, x and y both must be odd and $x \equiv 3 \pmod 8$. We write

$$y^2 + 18 = x^3 + 8 = (x + 2) \cdot (x^2 - 2x + 4).$$

As $x \equiv 3 \pmod 8$, $x + 2 \equiv 5 \pmod 8$. We need to discuss two cases.

Case I $x + 2$ has a prime divisor $p \equiv 5 \pmod 8$. In this case, we have the following contradiction:

$$1 = \left(\frac{-18}{p}\right) = \left(\frac{2}{p}\right) = -1.$$

Case II $x + 2$ has a prime divisor $p \equiv 3 \pmod 8$ and a prime divisor $q \equiv 7 \pmod 8$. In this case, we have the following contradiction:

$$1 = \left(\frac{-18}{q}\right) = -\left(\frac{2}{q}\right) = -1.$$

CHAPTER 12

Some Additional Topics

1. Hensel's Lemma

To motivate the discussion about Hensel's lemma, let us try solving the congruence $x^2 \equiv 2 \pmod{7^4}$. We note that

$$x^2 \equiv 2 \pmod{7^4} \Longrightarrow x^2 \equiv 2 \pmod{7^3} \Longrightarrow x^2 \equiv 2 \pmod{7^2} \Longrightarrow x^2 \equiv 2 \pmod{7}.$$

The last congruence has a solution, namely, $x \equiv \pm 3 \pmod{7}$. Hensel's lemma deals with the problem of "lifting" a solution such as $x \equiv 3 \pmod{7}$ of the congruence $x^2 \equiv 2 \pmod{7}$ to solutions of the congruence $x^2 \equiv 2 \pmod{7^k}$ for $k > 1$. To do this by the brute force method, we proceed as follows. $x \equiv 3 \pmod{7} \Longrightarrow x = 3 + 7r$. Substituting this in $x^2 \equiv 2 \pmod{7^2}$, we have

$$(3+7r)^2 \equiv 2 \pmod{7^2} \Longrightarrow 9 + 42r + 49r^2 \equiv 2 \pmod{7^2} \Longrightarrow 42r \equiv -7 \pmod{7^2}.$$

The last congruence gives $r \equiv 1 \pmod{7}$. Let $r = 1 + 7s$. So

$$x = 3 + 7r = 3 + 7(1 + 7s) = 10 + 7^2 s.$$

Substituting $x = 10 + 7^2 s$ in $x^2 \equiv 2 \pmod{7^3}$, we have

$$(10 + 7^2 s)^2 \equiv 2 \pmod{7^3} \Longrightarrow 100 + 20 \cdot 7^2 s + 7^4 s^2 \equiv 2 \pmod{7^3} \Longrightarrow s \equiv 2 \pmod{7}.$$

Let $s = 2 + 7t$. Substituting this in $x = 10 + 7^2 s$, we get

$$x = 10 + 7^2(2 + 7t) = 108 + 7^3 t.$$

Finally, substituting the last value of x in $x^2 \equiv 2 \pmod{7^4}$ and simplifying, we get $t \equiv 6 \pmod{7}$. Let $t = 6 + 7k$. Substituting this in $x = 108 + 7^3 t$, we get $\mathbf{x \equiv 2166 \pmod{7^4}}$. This method of solving the congruence can be considerably simplified. We recall that in a beginning calculus course students learn Newton Raphson method of finding approximate value of x satisfying $f(x) = 0$, where $f(x)$ is some differentiable function of x. In this method, we start with some suitable "guess", say x_1, which is such that $f'(x_1) \neq 0$, as an approximate solution of $f(x) = 0$ and find the point on the x-axis where the tangent to the graph $y = f(x)$ at the point (x_1, y_1), where $y_1 = f(x_1)$ intersects the x-axis. Generally, this point will be a better approximation for a solution for $f(x) = 0$ than x_1. We name this new point x_2 and repeat this process. The equation of the tangent line to the graph of $y = f(x)$ at the point (x_1, y_1) is

$$Y - f(x_1) = f^{'}(x_1)(X - x_1).$$

This tangent line intersects the x-axis at the point whose x-coordinate is $x_2 = x_1 - \dfrac{f(x_1)}{f'(x_1)}$, or $x_2 = x_1 - \left(f^{'}(x_1)^{-1}\right)f(x_1)$. Generally, x_2 will be a better approximation to a root of $f(x) = 0$ than x_1 as indicated in the accompanying diagram.

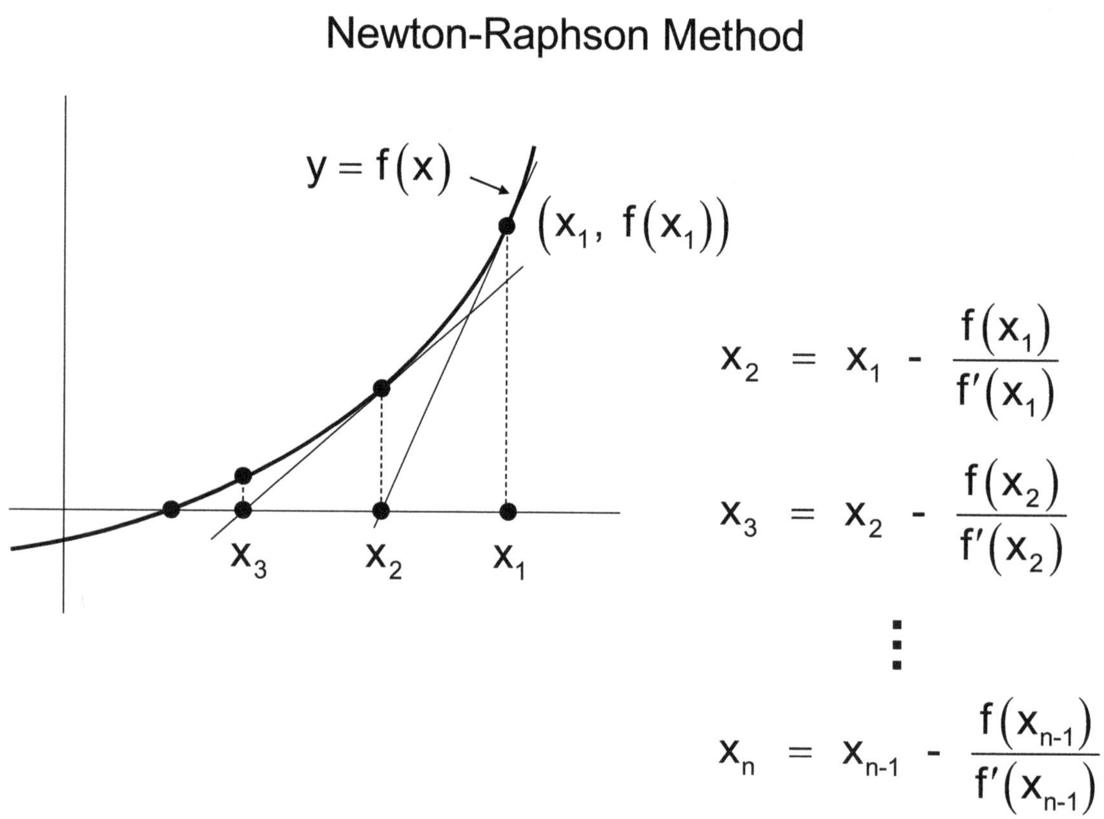

Repeating this process, we at the nth iteration get the following approximation to a solution of $f(x) = 0$:

$$x_n = x_{n-1} - f^{'}(x_{n-1})^{-1}f(x_{n-1}).$$

The procedure of Hensel's lemma for solving the polynomial congruence $f(x) \equiv 0 \pmod{p^n}$, where p is a prime number is somewhat analogous to the Newton Raphson method described above. In Hensel's lemma, we first find a solution $x = x_1$ to the congruence $f(x) \equiv 0 \pmod{p}$. Assuming that $f'(x_1) \not\equiv 0 \pmod{p}$, we find $b \in \mathbb{N}$ such that $b \cdot f'(x_1) \equiv 1 \pmod{p}$ or $b \equiv \dfrac{1}{f'(x_1)} = f'(x_1)^{-1} \pmod{p}$. We let

$$x_2 = x_1 - \frac{f(x_1)}{f'(x_1)} \equiv x_1 - b \cdot f(x_1) \pmod{p^2}.$$

Continue this we will have at the nth step

$$x_n = x_{n-1} - b \cdot f(x_{n-1}) \pmod{p^n}.$$

To apply this method to the congruence $x^2 \equiv 2 \pmod{7^4}$, we let $f(x) = x^2 - 2$, $x_1 = 3$, $f'(x_1) = 2x_1 = 6 \not\equiv 0 \pmod{7}$. We note that as $6^2 \equiv 1 \pmod{7}$, $6^{-1} \equiv 6 \pmod{7}$. Therefore, we let $b = \left(f'(x_1)\right)^{-1} = 6 \pmod{7}$.

$$\begin{aligned} x_2 &= x_1 - 6 \cdot f(x_1) = 3 - 42 = -39 \equiv 10 \pmod{7^2} \\ x_3 &= x_2 - 6 \cdot f(x_2) = 10 - 6 \cdot 98 = -578 \equiv 108 \pmod{7^3} \\ x_4 &= x_3 - 6 \cdot f(x_3) = 108 - 6 \cdot 11662 = -69864 \equiv 2111 \pmod{7^4} \end{aligned}$$

which is same as the solution we found by brute force method earlier. We summarize the description of the method of solving the congruence $x^2 \equiv 2 \pmod{7^4}$ in the following table. Let $f(x) = x^2 - 2$, $x_1 = 3$, $f'(x_1) = 2x_1 = 6$, $f'(x_1)^{-1} = 6^{-1} \equiv 6 \pmod{7}$. The following tables show the results of extending the above solutions of $f(x) \equiv 0 \pmod{7}$ to the solutions of $f(x) \equiv 0 \pmod{7^4}$, for $2 \leq i \leq 4$.

$$x_1 = 3, \qquad f'(3) = 6, \qquad 6^{-1} \equiv 6 \pmod{7}$$

i	x_i	$f(x_i) \pmod{7^{i+1}}$	$x_{i+1} = x_i - 6 \cdot f(x_i) \pmod{7^{i+1}}$
1	3	7	$3 - 6 \cdot 7 = -39 \equiv 10 \pmod{7^2}$
2	10	98	$10 - 6 \cdot 98 = -578 \equiv 108 \pmod{7^3}$
3	108	$11662 \equiv 2058$	$108 - 6 \cdot 2058 = -12240 \equiv 2166 \pmod{7^4}$

Another Example

Let us solve the congruence $f(x) = x^2 + x + 10 \equiv 0 \pmod{11^4}$ using Hensel's lemma. To do this, we first solve $f(x) \equiv 0 \pmod{11}$. The two solutions of the latter congruence are $x \equiv 3, 7 \pmod{11}$. As $f'(x) = 2x + 1$, $f'(3) = 7 \not\equiv 0 \pmod{11}$, and $f'(7) = 15 \not\equiv 0 \pmod{11}$. The following tables show the results of extending the above solutions of $f(x) \equiv 0 \pmod{11}$ to the solutions of $f(x) \equiv 0 \pmod{11^i}$, for $2 \le i \le 4$.

$$x_1 = 3, \qquad f'(3) = 7, \qquad 7^{-1} \equiv 8 \pmod{11}$$

i	x_i	$f(x_i) \pmod{11^{i+1}}$	$x_{i+1} = x_i - 8 \cdot f(x_i) \pmod{11^{i+1}}$
1	3	22	$3 - 8 \cdot 22 = -173 \equiv 69$
2	69	$4840 \equiv 847$	$69 - 8 \cdot 847 = -6707 \equiv 1279$
3	1279	$1637130 \equiv 11979$	$1279 - 8 \cdot 11979 = -94553 \equiv 7934$

$$x_1 = 7, \qquad f'(7) = 15 \equiv 4 \pmod{11}, \qquad 4^{-1} \equiv 3 \pmod{11}$$

i	x_i	$f(x_i) \pmod{11^{i+1}}$	$x_{i+1} = x_i - 3 \cdot f(x_i) \pmod{11^{i+1}}$
1	7	66	$7 - 3 \cdot 66 = -191 \equiv 51$
2	51	$2662 \equiv 0$	$51 - 3 \cdot 0 = 51$
3	51	2662	$51 - 3 \cdot 2662 = -7935 \equiv 6706$

So the two solutions to $f(x) \equiv 0 \pmod{11^4}$ are $x \equiv 7934$ or $6706 \pmod{11^4}$.

2. LCM(a, b, c) and GCD(a, b, c)

Relationship between LCM(a, b, c) and GCD(a, b, c)

Prove that if a, b, c are three natural numbers then

1. $\text{LCM}(a, b, c) \times \text{GCD}(ab, bc, ca) = a \cdot b \cdot c$.
2. $\text{GCD}(a, b, c) \times \text{LCM}(ab, bc, ca) = a \cdot b \cdot c$.

Proof:

It can be easily checked that if x, y, z are three real numbers then

- $\max(x, y, z) + \min(x + y, y + z, z + x) = x + y + z$
- $\min(x, y, z) + \max(x + y, y + z, z + x) = x + y + z$

Assume that a, b, and c are expressed as a product of primes as follows:

$$\begin{aligned} a &= p_1^{l_1}p_2^{l_2}p_3^{l_3}\cdots p_k^{l_k} \\ b &= p_1^{m_1}p_2^{m_2}p_3^{m_3}\cdots p_k^{m_k} \\ c &= p_1^{n_1}p_2^{n_2}p_3^{n_3}\cdots p_k^{n_k}. \end{aligned}$$

We know that

$$\begin{aligned} \text{LCM}(a,b,c) &= p_1^{\max(l_1,m_1,n_1)}p_2^{\max(l_2,m_2,n_2)}\cdots p_k^{\max(l_k,m_k,n_k)} \\ \text{GCD}(a,b,c) &= p_1^{\min(l_1,m_1,n_1)}p_2^{\min(l_2,m_2,n_2)}\cdots p_k^{\min(l_k,m_k,n_k)} \\ \text{GCD}(ab,bc,ca) &= p_1^{\min(l_1+m_1,m_1+n_1,n_1+l_1)}p_2^{\min(l_2+m_2,m_2+n_2,n_2+l_2)}\cdots p_k^{\min(l_k+m_k,m_k+n_k,n_k+l_k)} \\ \text{LCM}(ab,bc,ca) &= p_1^{\max(l_1+m_1,m_1+n_1,n_1+l_1)}p_2^{\max(l_2+m_2,m_2+n_2,n_2+l_2)}\cdots p_k^{\max(l_k+m_k,m_k+n_k,n_k+l_k)} \end{aligned}$$

Now the product LCM(a,b,c) $\times$ GCD(ab,bc,ca) is equal to the product of the following k prime powers:

$$\begin{gathered} p_1^{\max(l_1,m_1,n_1)+\min(l_1+m_1,m_1+n_1,n_1+l_1)} \\ p_2^{\max(l_2,m_2,n_2)+\min(l_2+m_2,m_2+n_2,n_2+l_2)} \\ p_3^{\max(l_3,m_3,n_3)+\min(l_3+m_3,m_3+n_3,n_3+l_3)} \\ \cdots\cdots\cdots \\ p_k^{\max(l_k,m_k,n_k)+\min(l_k+m_k,m_k+n_k,n_k+l_k)} \end{gathered}$$

which is same as

$$p_1^{l_1+m_1+n_1}p_2^{l_2+m_2+n_2}p_3^{l_3+m_3+n_3}\cdots p_k^{l_k+m_k+n_k} = a\cdot b\cdot c.$$

This completes the proof of the first statement. The proof of the second is similar.

An Example:

Let $a = 840 = 2^3\cdot 3\cdot 5\cdot 7,\ \ b = 2700 = 2^2\cdot 3^3\cdot 5^2,\ \ c = 2250 = 2\cdot 3^2\cdot 5^3$

a	b	c	ab	bc	ca	abc	LCM(a,b,c)	GCD(a,b,c)
$2^3\cdot 3\cdot 5\cdot 7$	$2^2 3^3 5^2$	$2^1 3^2 5^3$	$2^5 3^4 5^3 7$	$2^3 3^5 5^5$	$2^4 3^3 5^4 7$	$\mathbf{2^6 3^6 5^6 7}$	$2^3 3^3 5^3 7$	$2\cdot 3\cdot 5$

$$\begin{aligned} \text{LCM}(ab,bc,ca) &= 2^5\cdot 3^5\cdot 5^5\cdot 7 \\ \text{GCD}(ab,bc,ca) &= 2^3\cdot 3^3\cdot 5^3 \end{aligned}$$

$$\begin{aligned} \text{LCM}(a,b,c)\times\text{GCD}(ab,bc,ca) &= 2^3\cdot 3^3\cdot 5^3\cdot 7\times 2^3\cdot 3^3\cdot 5^3 = 2^6\cdot 3^6\cdot 5^6\cdot 7 = a\cdot b\cdot c. \\ \text{GCD}(a,b,c)\times\text{LCM}(ab,bc,ca) &= 2\cdot 3\cdot 5\times 2^5\cdot 3^5\cdot 5^5\cdot 7 \quad = 2^6\cdot 3^6\cdot 5^6\cdot 7 = a\cdot b\cdot c. \end{aligned}$$

We now generalize the previous theorem to the relationship between the LCM and GCD of n numbers.

Theorem: Let x_i, $1 \leq i \leq r$ be natural numbers and $N = x_1 x_2 \cdot x_2 \cdots x_r$ be their product. Then

$$\text{LCM}(x_1, x_2, \cdots x_r) \times \text{GCD}\left(\frac{N}{x_1}, \frac{N}{x_2}, \cdots, \frac{N}{x_r}\right) = N$$

and

$$\text{GCD}(x_1, x_2, \cdots x_r) \times \text{LCM}\left(\frac{N}{x_1}, \frac{N}{x_2}, \cdots, \frac{N}{x_r}\right) = N$$

Lemma:

Let a_i, $1 \leq i \leq n$ be any real numbers and $A = a_1 + a_2 + \cdots + a_n$. Then

$$\max(a_1, a_2, \cdots, a_n) + \min(A - a_1, A - a_2, \cdots, A - a_n) = A$$

Proof: Immediate consequence of the definition of maximum and minimum.

Proof of the Theorem:

Let

$$\begin{aligned} x_1 &= p_1^{a_{11}} p_2^{a_{12}} \cdots p_k^{a_{1k}} \\ x_2 &= p_1^{a_{21}} p_2^{a_{22}} \cdots p_k^{a_{2k}} \\ &\vdots \\ x_r &= p_1^{a_{r1}} p_2^{a_{r2}} \cdots p_k^{a_{rk}} \end{aligned}$$

Let

$$\begin{aligned} A_1 &= a_{11} + a_{21} + \cdots + a_{r1} \\ A_2 &= a_{12} + a_{22} + \cdots + a_{r2} \\ &\vdots \\ A_k &= a_{1k} + a_{2k} + \cdots + a_{rk} \end{aligned}$$

Therefore

$$\begin{aligned} \frac{N}{x_1} &= p_1^{A_1 - a_{11}} \cdot p_2^{A_2 - a_{12}} \cdots p_k^{a_{1k}} \\ \frac{N}{x_2} &= p_1^{A_1 - a_{12}} \cdot p_2^{A_2 - a_{22}} \cdots p_k^{a_{2k}} \\ &\vdots \\ \frac{N}{x_r} &= p_1^{A_1 - a_{r1}} \cdot p_2^{A_2 - a_{r2}} \cdots p_k^{a_{rk}} \end{aligned}$$

Then

$$\text{LCM}(x_1,x_2,\cdots,x_r)=p_1^{\max(a_{11},a_{21},\cdots,a_{r1})}\cdot p_2^{\max(a_{12},a_{22},\cdots,a_{r2})}\cdots p_k^{\max(a_{1k},a_{2k},\cdots,a_{rk})}$$

$$\text{GCD}(x_1,x_2,\cdots,x_r)=p_1^{\min(a_{11},a_{21},\cdots,a_{r1})}\cdots p_2^{\min(a_{12},a_{22},\cdots,a_{r2})}\cdots p_k^{\min(a_{1k},a_{2k},\cdots,a_{rk})}$$

$$\text{GCD}\left(\frac{N}{x_1},\frac{N}{x_2},\cdots,\frac{N}{x_r}\right)=\prod_{i=1}^{i=k}p_i^{\min(A_i-a_{1i},A_i-a_{2i},\cdots,A_i-a_{ri})}$$

$$\text{LCM}\left(\frac{N}{x_1},\frac{N}{x_2},\cdots,\frac{N}{x_r}\right)=\prod_{i=1}^{i=k}p_i^{\max(A_i-a_{1i},A_i-a_{2i},\cdots,A_i-a_{ri})}$$

As we noted before for $1\le i\le k$, we have

$$\max(a_{1i},a_{2i},\cdots,a_{ri})+\min(A_i-a_{1i},A_i-a_{2i},\cdots,A_i-a_{r1})=A_i$$

Finally, we have

$$\text{LCM}(x_1,x_2,\cdots x_r)\times\text{GCD}\left(\frac{N}{x_1},\frac{N}{x_2},\cdots,\frac{N}{x_r}\right)=p_1^{A_1}\cdot p_2^{A_2}\cdots p_k^{A_k}=x_1\cdot x_2\cdots x_k=N.$$

This proves the first part. The proof of the second part is similar.

3. Orders of $GL(k,\mathbb{Z}_n)$ and $SL(k,\mathbb{Z}_n)$

Calculating Order of $GL(k,\mathbb{Z}_n)$ and $SL(k,\mathbb{Z}_n)$

Let $GL(k,\mathbb{Z}_n)$ denote the set of all $k\times k$ matrices whose determinant is a unit in $\mathbb{Z}_n$. So if $A\in GL(k,\mathbb{Z}_n)$ then its entries belong to the set $\{0,1,2,\cdots,n-1\}$ and its determinant is co-prime to n. It is easily seen that $GL(k,\mathbb{Z}_n)$ is a group. $SL(k,\mathbb{Z}_n)$ is a subgroup of $GL(k,\mathbb{Z}_n)$ consisting of those element of $GL(k,\mathbb{Z}_n)$ whose determinant is 1. It is easily seen that if $n=p$, a prime number then

$$|GL(k,\mathbb{Z}_p)|=(p^k-1)(p^k-p)(p^k-p^2)\cdots\cdot(p^k-p^{k-1})=\prod_{i=0}^{k-1}(p^k-p^i)$$

and

$$|SL(k,\mathbb{Z}_p)|=\frac{1}{p-1}(p^k-1)(p^k-p)(p^k-p^2)\cdots\cdot(p^k-p^{k-1})=\frac{1}{p-1}\prod_{i=0}^{k-1}(p^k-p^i)$$

$$=p^{k-1}\prod_{i=0}^{k-2}(p^k-p^i).$$

We calculate the order of $GL(k,\mathbb{Z}_n)$ using the Chinese Remainder Theorem. Let $n = p_1^{a_1}p_2^{a_2}\cdots p_r^{a_r}$. Consider the mapping

$$GL(k,\mathbb{Z}_n) \longrightarrow GL(k,\mathbb{Z}_{p_1^{a_1}}) \times GL(k,\mathbb{Z}_{p_2^{a_2}}) \times \cdots\cdots GL(k,\mathbb{Z}_{p_r^{a_r}})$$

which maps $A \in GL(k,\mathbb{Z}_n)$ to $(A_1, A_2, \cdots, A_r) \in GL(k,\mathbb{Z}_{p_1^{a_1}}) \times \cdots\cdots \times GL(k,\mathbb{Z}_{p_r^{a_r}})$ where $A_i \in GL(k,p_i^{a_i})$ is such that its entries are the entries of $A \in GL(k,\mathbb{Z}_n)$ reduced modulo $p_i^{a_i}$, $1 \le i \le r$. This mapping is easily seen to be one-to-one. The Chinese Remainder Theorem implies that this mapping is onto. Therefore

$$|GL(k,\mathbb{Z}_n)| = \prod_{i=1}^{i=r} |GL(k,p_i^{a_i})|.$$

This reduces the problem of computing the order of $GL(k,\mathbb{Z}_n)$ to the case when n is a power of a prime number. So we now let $n = p^a$, where p is a prime number. To do this consider the mapping

$$GL(k,\mathbb{Z}_{p^a}) \longrightarrow GL(k,\mathbb{Z}_p)$$

in which the image of $A \in GL(k,p^a)$ is obtained by reducing each of the entry of the matrix A modulo p. This mapping is a homomorphism whose kernel consists of those matrices $A \in GL(k,\mathbb{Z}_{p^a})$, whose diagonal entry is $\equiv 1 \pmod p$ and nondiagonal entry is $\equiv 0 \pmod p$. There are p^{a-1} choices for each entry of the matrix $A \in GL(k,\mathbb{Z}_{p^a})$ which is in the kernel of the above mapping. Since there are k^2 entries in A, the order of the kernel is $p^{k^2(a-1)}$. Since the above mapping is onto therefore the order of $GL(k,\mathbb{Z}_{p^a})$ is equal to the product of order of $GL(k,\mathbb{Z}_p$ by $p^{k^2(a-1)}$. Therefore

$$|GL(k,\mathbb{Z}_{p^a})| = p^{k^2(a-1)} \times |GL(k,\mathbb{Z}_p)| = p^{k^2(a-1)}(p^k-1)(p^k-p)(p^k-p^2).\cdots(p^k-p^{k-1}).$$

So if $n = p_1^{a_1}p_2^{a_2}\cdots p_r^{a_r}$ then

$$|GL(k,\mathbb{Z}_n) = \prod_{i=1}^{i=r} p_i^{k^2(a_i-1)}|GL(k,\mathbb{Z}_{p_i}|$$

Clearly $|GL(k,\mathbb{Z}_n)| = \phi(n)\cdot|SL(k,\mathbb{Z}_n)|$, we have

$$|SL(k,\mathbb{Z}_n)| = \frac{1}{\phi(n)}\prod_{i=1}^{i=r} p_i^{k^2(a_i-1)}|GL(k,\mathbb{Z}_{p_i}|.$$

Substituting $k = 2$, we have

$$\begin{aligned}
|GL(2,\mathbb{Z}_n)| &= \prod_{i=1}^{i=r} p_i^{4(a_i-1)}|GL(2,\mathbb{Z}_{p_i}| \\
&= \prod_{i=1}^{i=r} \frac{p_i^{4_i}}{p_i^4}(p_i^2-1)(p_i^2-p_i) \\
&= n^4 \prod_{i=1}^{i=r}\left(1-\frac{1}{p_i^2}\right)\left(1-\frac{1}{p_i}\right) \\
&= \phi(n)n^3 \prod_{i=1}^{i=r}\left(1-\frac{1}{p_i^2}\right) \\
&= \phi(n)n^3 \prod_{p|n}\left(1-\frac{1}{p^2}\right)
\end{aligned}$$

Therefore

$$|SL(2,\mathbb{Z}_n)| = \frac{1}{\phi(n)}|GL(2,\mathbb{Z}_n)| = n^3 \prod_{p|n}\left(1-\frac{1}{p^2}\right).$$

In particular $|GL(2,4)| = 96$, and $|SL(2,4)| = 48$ as can be easily verified.

4. Problem of Five Sailors and a Monkey

In an article published in *Mathematical Intelligencer*, Halmos (1991) states the following problem concerning five sailors and a monkey:

> After gathering a pile of coconuts one day, five sailors on a desert island agree to divide them evenly next morning. During the night one sailor secretly gets up and tries to divide the coconuts into five equal piles. Finding that there is a remainder of one coconut, he tosses the odd one to a monkey, and, secreting his pile, mixes up the remaining, and goes back to sleep. The second sailor does the same thing, with the same result, and so do the third, fourth, and fifth. In the morning the remaining pile of coconuts (less one) is again divisible by 5. What is the smallest number of coconuts that the original pile contained?

Halmos gives a solution (attributed to P. A. M. Dirac) which is as follows:

For each possible number x of coconuts, let $S(x)$ denote the number of coconuts left by a sailor's (dishonest) activity. Then clearly

$$S(x) = x - \frac{x-1}{5} - 1 = \frac{4}{5}(x-1) = \frac{4}{5}x - \frac{4}{5}.$$

By considering S as an operator and applying it six times on two piles of coconuts, respectively, containing x and y coconuts it follows that:

$$S^6(x) = \left(\frac{4}{5}\right)^6 x - \left(\frac{4}{5}\right)^5 - \left(\frac{4}{5}\right)^4 - \left(\frac{4}{5}\right)^3 - \left(\frac{4}{5}\right)^2 - \frac{4}{5}$$

$$S^6(y) = \left(\frac{4}{5}\right)^6 y - \left(\frac{4}{5}\right)^5 - \left(\frac{4}{5}\right)^4 - \left(\frac{4}{5}\right)^3 - \left(\frac{4}{5}\right)^2 - \frac{4}{5}$$

So

$$S^6(x) - S^6(y) = \left(\frac{4}{5}\right)^6 (x-y).$$

He concludes that if both x and y are solutions, then x and y are congruent modulo 5^6. The simplest solution corresponds to $S(x) = x$, namely $\frac{4}{5}(x-1) = x$, i.e., for $x = -4$. So any arbitrary solution is given by $k5^6 - 4$, for $k \geq 1$.

The purpose of this note is to generalize this problem when there are n sailors and r monkeys. In the generalization, we change the problem slightly and assume that each of the sailors first tosses a coconut to each of the r monkeys and finds that the remainder is evenly divisible by n. Starting with x coconuts, the value of $S(x)$ is now given by

$$S(x) = x - r - \frac{x-r}{n} = \frac{n-1}{n}x - \frac{n-1}{n}r.$$

Applying this process $n+1$ times, we get

$$S^{n+1}(x) = \left(\frac{n-1}{n}\right)^{n+1} x - \left[\left(\frac{n-1}{n}\right)^{n+1} + \cdots + \left(\frac{n-1}{n}\right)\right] r.$$

It follows that if we start with two piles of coconuts containing x and y coconuts, respectively, then:

$$S^{n+1}(x) - S^{n+1}(y) = \left(\frac{n-1}{n}\right)^{n+1} (x-y).$$

Therefore , $x \equiv y \pmod{n^{n+1}}$. The simplest solution corresponds to $S(x) = x$ which gives

$$\frac{n-1}{n}x - \frac{n-1}{n}r = x, \qquad \text{or} \qquad x = -(n-1)r$$

so a general solution is given by $kn^{n+1} - (n-1)r$, where $k > 0$, is an integer and is such that this expression is positive. The following table gives the least positive solution for various values of n and r, together with the value of n^{n+1}—the difference between two successive solutions.

Number of Sailors (n)	Number of Monkeys	Minimal Positive Solution	n^{n+1}
3	1	79	81
3	2	77	81
4	1	1,021	1,024
4	2	1,018	1,024
4	3	1,015	1,024
5	1	15,621	15,625
5	2	15,617	15,625
5	3	15,613	15,625
5	4	15,609	15,625
6	1	279,931	279,936

Appendix 1

First 12 Mersenne Primes

N	p	# Digits in $2^p - 1$	N-th Mersenne Prime $2^p - 1$
1	2	1	3
2	3	1	7
3	5	2	31
4	7	3	127
5	13	4	8191
6	17	6	131071
7	19	6	524287
8	31	10	2147483647
9	61	19	2305843009213693951
10	89	27	618970019642690137449562111
11	107	33	162259276829213363391578010288127
12	127	39	170141183460469231731687303715884105727

Appendix 2

Algebraic Numbers and Algebraic Integers

Algebraic Numbers

A complex number α is said to be an *algebraic number* if there exists a nonzero polynomial $f(x) \in \mathbb{Q}[x]$ such that $f(\alpha) = 0$.

If α is an algebraic number then there exists a unique nonzero polynomial $f_\alpha(x) \in \mathbb{Q}[x]$ satisfying the following conditions:

(i) $f_\alpha(x)$ is monic and $f_\alpha(\alpha) = 0$.

(ii) If $g(x) \in \mathbb{Q}[x]$ is any polynomial such that $g(\alpha) = 0$, then degree $f_\alpha(x) \leq$ degree $g(x)$.

Such a (uniquely determined) polynomial is called the minimal polynomial of the algebraic number α and its degree is the degree of the algebraic number α.

The following table gives the minimal polynomial of various algebraic numbers.

Algebraic Numberα	Minimal Polynomial$f_\alpha(x)$
$\sqrt{2}$	$x^2 - 2$
$\frac{1}{\sqrt{2}}$	$x^2 - \frac{1}{2}$
$\sqrt{2} + \sqrt{3}$	$x^4 - 10x^2 + 19$
$\frac{1+i}{\sqrt{2}}$	$x^4 + 1$
$\cos\frac{2\pi}{9} + i\sin\frac{2\pi}{9}$	$x^6 + x^3 + 1$

Algebraic Integers

A complex number α is said to be an *algebraic integer* if there exists a monic (that is its leading coefficient is unit) nonzero polynomial $f(x) \in Z[x]$ such that $f(\alpha) = 0$.

Evidently an algebraic integer is also an algebraic number and its minimal polynomial and degree are similarly defined. In the table above all algebraic numbers excepting $\frac{1}{\sqrt{2}}$ are algebraic integers.

Algebraic Numbers and Algebraic Integers

We shall now prove that the set of all algebraic numbers form a field and the set of all algebraic integers form a ring.

Theorem If $\alpha \neq 0$ is an algebraic number, then so is $\frac{1}{\alpha}$. If α and β are algebraic numbers then so are $\alpha + \beta$ and $\alpha \cdot \beta$. In other words the set of all algebraic numbers form a field.

Proof

If $\alpha \neq 0$ is a zero of the polynomial $f(x) = \sum_{i=0}^{n} a_i x^i \in \mathbb{Q}[x]$ then $\frac{1}{\alpha}$ is a zero of the polynomial $g(x) = \sum_{i=0}^{n} a_{n-i} x^i \in \mathbb{Q}[x]$. Therefore $\frac{1}{\alpha}$ is an algebraic number.

Let degree $\alpha = m$ and degree $\beta = n$. Let $f(x)$ and $g(x)$ be respectively the minimal polynomials of the algebraic numbers α and β. So $f(x)$ and $g(x)$ are monic polynomials with rational coefficients and degree m and n respectively. Consider the mn numbers $\alpha^i \beta^j$, where $0 \leq i \leq m-1$, $0 \leq j \leq n-1$. Let γ_k, $1 \leq k \leq mn$ be some enumeration of these mn numbers. Let $mn = N$. With the help of the polynomials $f(x)$ and $g(x)$ we can express α^m as a linear combination of lower powers of α with rational coefficients β^n as a linear combination of lower powers of β with rational coefficients. Therefore we can write

$$(\alpha + \beta)\gamma_k = \sum_{\ell=1}^{N} a_{k\ell}\gamma_\ell, \qquad a_{k\ell} \in \mathbb{Q}$$

Likewise

$$(\alpha \cdot \beta)\gamma_k = \sum_{\ell=1}^{N} b_{k\ell}\gamma_\ell, \qquad b_{k\ell} \in \mathbb{Q}$$

So we can think of $\alpha + \beta$ as an eigenvalue of the $N \times N$ rational matrix $[a_{k\ell}]$ and $\alpha \cdot \beta$ as an eigenvalue of the $N \times N$ rational matrix $[b_{k\ell}]$.

Clearly the eigenvalues of a (square) rational matrix are algebraic numbers (since the characteristic polynomial of such a matrix has rational coefficients). It follows that both $\alpha + \beta$ and $\alpha\beta$ are algebraic numbers. Since the set of algebraic numbers is a subset of the set of complex numbers the other field axioms are easily seen to be valid. Hence the set of algebraic numbers form a field.

Theorem The set of all algebraic integers form a ring.

Proof This is proved exactly as in the preceding theorem. The matrices $[a_{k\ell}]$ and $[b_{k\ell}]$ in this case will be integral matrices and so their eigen values are algebraic integers.

Note that the reciprocal of an algebraic integer may not be an algebraic integer as 2 is an algebraic integer but $\frac{1}{2}$ is not.

Appendix 3

Continued Fraction for square root of n for $2 < n < 500$ (non-square n)

2	1 2
3	1 1 2
5	2 4
6	2 2 4
7	2 1 1 1 4
8	2 1 4
10	3 6
11	3 3 6
12	3 2 6
13	3 1 1 1 1 6
14	3 1 2 1 6
15	3 1 6
17	4 8
18	4 4 8
19	4 2 1 3 1 2 8
20	4 2 8
21	4 1 1 2 1 1 8
22	4 1 2 4 2 1 8
23	4 1 3 1 8
24	4 1 8

Continued Fraction for square root of n for $2 < n < 500$ (non-square n)

26	5	10											
27	5	5	10										
28	5	3	2	3	10								
29	5	2	1	1	2	10							
30	5	2	10										
31	5	1	1	3	5	3	1	1	10				
32	5	1	1	1	10								
33	5	1	2	1	10								
34	5	1	4	1	10								
35	5	1	10										
37	6	12											
38	6	6	12										
39	6	4	12										
40	6	3	12										
41	6	2	2	12									
42	6	2	12										
43	6	1	1	3	1	5	1	3	1	1	12		
44	6	1	1	1	2	1	1	1	12				
45	6	1	2	2	2	1	12						
46	6	1	3	1	1	2	6	2	1	1	3	1	12
47	6	1	5	1	12								
48	6	1	12										
50	7	14											

Continued Fraction for square root of n for $2 < n < 500$ (non-square n)

51	7	7	14									
52	7	4	1	2	1	4	14					
53	7	3	1	1	3	14						
54	7	2	1	6	1	2	14					
55	7	2	2	2	14							
56	7	2	14									
57	7	1	1	4	1	1	14					
58	7	1	1	1	1	1	1	14				
59	7	1	2	7	2	1	14					
60	7	1	2	1	14							
61	7	1	4	3	1	2	2	1	3	4	1	14
62	7	1	6	1	14							
63	7	1	14									
65	8	16										
66	8	8	16									
67	8	5	2	1	1	7	1	1	2	5	16	
68	8	4	16									
69	8	3	3	1	4	1	3	3	16			
70	8	2	1	2	1	2	16					
71	8	2	2	1	7	1	2	2	16			
72	8	2	16									
73	8	1	1	5	5	1	1	16				
74	8	1	1	1	1	16						
75	8	1	1	1	16							

Continued Fraction for square root of n for $2 < n < 500$ (non-square n)

76	8	1	2	1	1	5	4	5	1	1	2	1	16				
77	8	1	3	2	3	1	16										
78	8	1	4	1	16												
79	8	1	7	1	16												
80	8	1	16														
82	9	18															
83	9	9	18														
84	9	6	18														
85	9	4	1	1	4	18											
86	9	3	1	1	1	8	1	1	1	3	18						
87	9	3	18														
88	9	2	1	1	1	2	18										
89	9	2	3	3	2	18											
90	9	2	18														
91	9	1	1	5	1	5	1	1	18								
92	9	1	1	2	4	2	1	1	18								
93	9	1	1	1	4	6	4	1	1	1	18						
94	9	1	2	3	1	1	5	1	8	1	5	1	1	3	2	1	18
95	9	1	2	1	18												
96	9	1	3	1	18												
97	9	1	5	1	1	1	1	1	1	5	1	18					
98	9	1	8	1	18												
99	9	1	18														

Continued Fraction for square root of n for $2 < n < 500$ (non-square n)

101	10 20
102	10 10 20
103	10 6 1 2 1 1 9 1 1 2 1 6 20
104	10 5 20
105	10 4 20
106	10 3 2 1 1 1 1 2 3 20
107	10 2 1 9 1 2 20
108	10 2 1 1 4 1 1 2 20
109	10 2 3 1 2 4 1 6 6 1 4 2 1 3 2 20
110	10 2 20
111	10 1 1 6 1 1 20
112	10 1 1 2 1 1 20
113	10 1 1 1 2 2 1 1 1 20
114	10 1 2 10 2 1 20
115	10 1 2 1 1 1 1 1 2 1 20
116	10 1 3 2 1 4 1 2 3 1 20
117	10 1 4 2 4 1 20
118	10 1 6 3 2 10 2 3 6 1 20
119	10 1 9 1 20
120	10 1 20
122	11 22
123	11 11 22
124	11 7 2 1 1 1 3 1 4 1 3 1 1 1 2 7 22
125	11 5 1 1 5 22

Continued Fraction for square root of n for $2 < n < 500$ (non-square n)

126	11	4	2	4	22														
127	11	3	1	2	2	7	11	7	2	2	1	3	22						
128	11	3	5	3	22														
129	11	2	1	3	1	6	1	3	1	2	22								
130	11	2	2	22															
131	11	2	4	11	4	2	22												
132	11	2	22																
133	11	1	1	7	5	1	1	1	2	1	1	1	5	7	1	1	22		
134	11	1	1	2	1	3	1	10	1	3	1	2	1	1	22				
135	11	1	1	1	1	1	1	1	22										
136	11	1	1	1	22														
137	11	1	2	2	1	1	2	2	1	22									
138	11	1	2	1	22														
139	11	1	3	1	3	7	1	1	2	11	2	1	1	7	3	1	3	1	22
140	11	1	4	1	22														
141	11	1	6	1	22														
142	11	1	10	1	22														
143	11	1	22																
145	12	24																	
146	12	12	24																
147	12	8	24																
148	12	6	24																
149	12	4	1	5	3	3	5	1	4	24									
150	12	4	24																

Continued Fraction for square root of n for $2 < n < 500$ (non-square n)

<table>
<tr><td>151</td><td>12 3 2 7 1 3 4 1 1 1 11 1 1 1 4 3 1 7 2 3 24</td></tr>
<tr><td>152</td><td>12 3 24</td></tr>
<tr><td>153</td><td>12 2 1 2 2 2 1 2 24</td></tr>
<tr><td>154</td><td>12 2 2 3 1 2 1 3 2 2 24</td></tr>
<tr><td>155</td><td>12 2 4 2 24</td></tr>
<tr><td>156</td><td>12 2 24</td></tr>
<tr><td>157</td><td>12 1 1 7 1 5 2 1 1 1 1 2 5 1 7 1 1 24</td></tr>
<tr><td>158</td><td>12 1 1 3 12 3 1 1 24</td></tr>
<tr><td>159</td><td>12 1 1 1 1 3 1 1 1 1 24</td></tr>
<tr><td>160</td><td>12 1 1 1 5 1 1 1 24</td></tr>
<tr><td>161</td><td>12 1 2 4 1 2 1 4 2 1 24</td></tr>
<tr><td>162</td><td>12 1 2 1 2 12 2 1 2 1 24</td></tr>
<tr><td>163</td><td>12 1 3 3 2 1 1 7 1 11 1 7 1 1 2 3 3 1 24</td></tr>
<tr><td>164</td><td>12 1 4 6 4 1 24</td></tr>
<tr><td>165</td><td>12 1 5 2 5 1 24</td></tr>
<tr><td>166</td><td>12 1 7 1 1 1 2 4 1 3 2 12 2 3 1 4 2 1 1 1 7 1 24</td></tr>
<tr><td>167</td><td>12 1 11 1 24</td></tr>
<tr><td>168</td><td>12 1 24</td></tr>
<tr><td>170</td><td>13 26</td></tr>
<tr><td>171</td><td>13 13 26</td></tr>
<tr><td>172</td><td>13 8 1 2 2 1 1 3 6 3 1 1 2 2 1 8 26</td></tr>
<tr><td>173</td><td>13 6 1 1 6 26</td></tr>
<tr><td>174</td><td>13 5 4 5 26</td></tr>
<tr><td>175</td><td>13 4 2 1 2 4 26</td></tr>
</table>

Continued Fraction for square root of n for $2 < n < 500$ (non-square n)

176	13 3 1 3 26
177	13 3 3 2 8 2 3 3 26
178	13 2 1 12 1 2 26
179	13 2 1 1 1 3 5 13 5 3 1 1 1 2 26
180	13 2 2 2 26
181	13 2 4 1 8 6 1 1 1 1 2 2 1 1 1 1 6 8 1 4 2 26
182	13 2 26
183	13 1 1 8 1 1 26
184	13 1 1 3 2 1 2 1 2 3 1 1 26
185	13 1 1 1 1 26
186	13 1 1 1 3 4 3 1 1 1 26
187	13 1 2 13 2 1 26
188	13 1 2 2 6 2 2 1 26
189	13 1 2 1 26
190	13 1 3 1 1 1 2 2 2 1 1 1 3 1 26
191	13 1 4 1 1 3 2 2 13 2 2 3 1 1 4 1 26
192	13 1 5 1 26
193	13 1 8 3 2 1 3 3 1 2 3 8 1 26
194	13 1 12 1 26
195	13 1 26
197	14 28
198	14 14 28
199	14 9 2 1 2 2 5 4 1 1 13 1 1 4 5 2 2 1 2 9 28
200	14 7 28

Continued Fraction for square root of n for $2 < n < 500$ (non-square n)

n	Continued fraction
201	14 5 1 1 1 2 1 8 1 2 1 1 1 5 28
202	14 4 1 2 2 1 4 28
203	14 4 28
204	14 3 1 1 6 1 1 3 28
205	14 3 6 1 4 1 6 3 28
206	14 2 1 5 14 5 1 2 28
207	14 2 1 1 2 1 1 2 28
208	14 2 2 1 2 2 28
209	14 2 5 3 2 3 5 2 28
210	14 2 28
211	14 1 1 9 5 1 2 2 1 1 4 3 1 13 1 3 4 1 1 2 2 1 5 9 1 1 28
212	14 1 1 3 1 1 1 6 1 1 1 3 1 1 28
213	14 1 1 2 6 1 8 1 6 2 1 1 28
214	14 1 1 1 2 3 1 4 9 1 1 5 3 14 3 5 1 1 9 4 1 3 2 1 1 1 28
215	14 1 1 1 28
216	14 1 2 3 2 1 28
217	14 1 2 1 2 1 1 9 4 9 1 1 2 1 2 1 28
218	14 1 3 3 1 28
219	14 1 3 1 28
220	14 1 4 1 28
221	14 1 6 2 6 1 28
222	14 1 8 1 28
223	14 1 13 1 28
224	14 1 28

Continued Fraction for square root of n for $2 < n < 500$ (non-square n)

n	Continued fraction
226	15 30
227	15 15 30
228	15 10 30
229	15 7 1 1 7 30
230	15 6 30
231	15 5 30
232	15 4 3 7 3 4 30
233	15 3 1 3 1 1 1 1 3 1 3 30
234	15 3 2 1 2 1 2 3 30
235	15 3 30
236	15 2 1 3 5 1 6 1 5 3 1 2 30
237	15 2 1 1 7 10 7 1 1 2 30
238	15 2 2 1 14 1 2 2 30
239	15 2 5 1 2 4 15 4 2 1 5 2 30
240	15 2 30
241	15 1 1 9 1 5 3 3 1 1 3 3 5 1 9 1 1 30
242	15 1 1 3 1 14 1 3 1 1 30
243	15 1 1 2 3 15 3 2 1 1 30
244	15 1 1 1 1 1 2 1 5 1 1 9 1 6 1 9 1 1 5 1 2 1 1 1 1 1 30
245	15 1 1 1 7 6 7 1 1 1 30
246	15 1 2 5 1 14 1 5 2 1 30
247	15 1 2 1 1 9 1 9 1 1 2 1 30
248	15 1 2 1 30
249	15 1 3 1 1 5 1 3 10 3 1 5 1 1 3 1 30
250	15 1 4 3 3 4 1 30

Continued Fraction for square root of n for $2 < n < 500$ (non-square n)

n	Continued fraction
251	15 1 5 2 1 2 2 15 2 2 1 2 5 1 30
252	15 1 6 1 30
253	15 1 9 1 1 1 2 1 7 4 2 2 2 4 7 1 2 1 1 1 9 1 30
254	15 1 14 1 30
255	15 1 30
257	16 32
258	16 16 32
259	16 10 1 2 3 4 3 2 1 10 32
260	16 8 32
261	16 6 2 3 7 1 3 1 2 1 3 1 7 3 2 6 32
262	16 5 2 1 2 1 10 16 10 1 2 1 2 5 32
263	16 4 1 1 1 1 15 1 1 1 1 4 32
264	16 4 32
265	16 3 1 1 2 2 1 1 3 32
266	16 3 4 3 32
267	16 2 1 15 1 2 32
268	16 2 1 2 3 3 1 3 1 10 8 10 1 3 1 3 3 2 1 2 32
269	16 2 2 32
270	16 2 3 6 3 2 32
271	16 2 6 10 1 4 1 1 2 1 2 1 15 1 2 1 2 1 1 4 1 10 6 2 32
272	16 2 32
273	16 1 1 10 1 1 32
274	16 1 1 4 4 1 1 32
275	16 1 1 2 1 1 32

Continued Fraction for square root of n for $2 < n < 500$ (non-square n)

276	16 1 1 1 1 2 2 2 1 1 1 1 32
277	16 1 1 1 4 10 1 7 2 2 3 3 2 2 7 1 10 4 1 1 1 32
278	16 1 2 16 2 1 32
279	16 1 2 2 1 2 2 1 32
280	16 1 2 1 2 1 32
281	16 1 3 4 1 1 6 6 1 1 4 3 1 32
282	16 1 3 1 4 1 3 1 32
283	16 1 4 1 1 1 3 10 1 15 1 10 3 1 1 1 4 1 32
284	16 1 5 1 3 2 1 4 8 4 1 2 3 1 5 1 32
285	16 1 7 2 7 1 32
286	16 1 10 3 3 2 3 3 10 1 32
287	16 1 15 1 32
288	16 1 32
290	17 34
291	17 17 34
292	17 11 2 1 3 8 3 1 2 11 34
293	17 8 1 1 8 34
294	17 6 1 4 1 6 34
295	17 5 1 2 3 2 6 2 3 2 1 5 34
296	17 4 1 7 1 4 34
297	17 4 3 1 1 2 1 1 3 4 34
298	17 3 1 4 5 1 1 5 4 1 3 34
299	17 3 2 3 34
300	17 3 8 3 34

Continued Fraction for square root of n for $2 < n < 500$ (non-square n)

301	17	2	1	6	3	1	2	2	1	1	8	11	2	4	2	11	8	1	1	2
	2	1	3	6	1	2	34													
302	17	2	1	1	1	4	2	1	16	1	2	4	1	1	1	2	34			
303	17	2	2	5	2	2	34													
304	17	2	3	2	1	1	1	1	1	2	3	2	34							
305	17	2	6	2	34															
306	17	2	34																	
307	17	1	1	11	5	1	3	17	3	1	5	11	1	1	34					
308	17	1	1	4	1	1	34													
309	17	1	1	2	1	2	4	1	1	1	8	6	1	10	1	6	8	1	1	1
	4	2	1	2	1	1	34													
310	17	1	1	1	1	5	3	1	2	1	3	5	1	1	1	1	34			
311	17	1	1	1	2	1	6	3	17	3	6	1	2	1	1	1	34			
312	17	1	1	1	34															
313	17	1	2	4	11	1	1	3	2	2	3	1	1	11	4	2	1	34		
314	17	1	2	1	1	2	1	34												
315	17	1	2	1	34															
316	17	1	3	2	8	2	3	1	34											
317	17	1	4	8	1	2	2	1	8	4	1	34								
318	17	1	4	1	34															
319	17	1	6	5	1	4	3	1	3	4	1	5	6	1	34					
320	17	1	7	1	34															
321	17	1	10	1	34															
322	17	1	16	1	34															
323	17	1	34																	
325	18	36																		

Continued Fraction for square root of n for $2 < n < 500$ (non-square n)

n	Continued fraction
326	18 18 36
327	18 12 36
328	18 9 36
329	18 7 4 2 1 1 4 1 1 2 4 7 36
330	18 6 36
331	18 5 5 1 6 2 3 1 1 2 1 2 1 11 2 1 1 17 1 1 2 11 1 2 1 2 1 1 3 2 6 1 5 5 36
332	18 4 1 1 8 1 1 4 36
333	18 4 36
334	18 3 1 1 1 2 5 1 2 2 11 1 3 7 18 7 3 1 11 2 2 1 5 2 1 1 1 3 36
335	18 3 3 3 36
336	18 3 36
337	18 2 1 3 1 11 2 4 1 3 3 1 4 2 11 1 3 1 2 36
338	18 2 1 1 2 36
339	18 2 2 2 1 17 1 2 2 2 36
340	18 2 3 1 1 1 1 8 1 1 1 1 3 2 36
341	18 2 6 1 8 2 1 2 1 2 8 1 6 2 36
342	18 2 36
343	18 1 1 11 1 5 3 1 17 1 3 5 1 11 1 1 36
344	18 1 1 4 1 3 1 4 1 1 36
345	18 1 1 2 1 6 1 2 1 1 36
346	18 1 1 1 1 36
347	18 1 1 1 2 4 1 17 1 4 2 1 1 1 36
348	18 1 1 1 8 1 1 1 36
349	18 1 2 7 7 2 1 36
350	18 1 2 2 2 1 36

Continued Fraction for square root of n for $2 < n < 500$ (non-square n)

351	18 1 2 1 3 2 2 2 3 1 2 1 36
352	18 1 3 5 9 5 3 1 36
353	18 1 3 1 2 1 1 1 1 1 1 2 1 3 1 36
354	18 1 4 2 2 18 2 2 4 1 36
355	18 1 5 3 3 1 6 1 3 3 5 1 36
356	18 1 6 1 1 2 1 8 1 2 1 1 6 1 36
357	18 1 8 2 8 1 36
358	18 1 11 1 1 1 3 1 1 4 1 5 2 18 2 5 1 4 1 1 3 1 1 1 11 1 36
359	18 1 17 1 36
360	18 1 36
362	19 38
363	19 19 38
364	19 12 1 2 3 1 8 1 3 2 1 12 38
365	19 9 1 1 9 38
366	19 7 1 1 1 2 12 2 1 1 1 7 38
367	19 6 2 1 3 1 1 2 1 12 19 12 1 2 1 1 3 1 2 6 38
368	19 5 2 5 38
369	19 4 1 3 2 7 4 7 2 3 1 4 38
370	19 4 4 38
371	19 3 1 4 1 3 38
372	19 3 2 12 2 3 38
373	19 3 5 5 3 38
374	19 2 1 18 1 2 38
375	19 2 1 2 1 5 1 2 1 2 38

Continued Fraction for square root of n for $2 < n < 500$ (non-square n)

376	19 2 1 1 3 1 2 2 4 2 2 1 3 1 1 2 38
377	19 2 2 2 38
378	19 2 3 1 4 1 3 2 38
379	19 2 7 3 2 2 6 12 1 4 1 1 1 3 4 19 4 3 1 1 1 4 1 12 6 2 2 3 7 2 38
380	19 2 38
381	19 1 1 12 1 1 38
382	19 1 1 5 12 1 5 1 1 2 3 1 18 1 3 2 1 1 5 1 12 5 1 1 38
383	19 1 1 3 19 3 1 1 38
384	19 1 1 2 9 2 1 1 38
385	19 1 1 1 1 1 3 1 2 1 3 1 1 1 1 1 38
386	19 1 1 1 4 1 18 1 4 1 1 1 38
387	19 1 2 19 2 1 38
388	19 1 2 3 4 12 1 8 1 12 4 3 2 1 38
389	19 1 2 1 1 1 1 2 1 38
390	19 1 2 1 38
391	19 1 3 2 2 1 1 2 19 2 1 1 2 2 3 1 38
392	19 1 3 1 38
393	19 1 4 1 2 4 1 1 1 1 12 1 1 1 1 4 2 1 4 1 38
394	19 1 5 1 1 1 3 1 3 5 2 2 5 3 1 3 1 1 1 5 1 38
395	19 1 6 1 38
396	19 1 8 1 38
397	19 1 12 3 4 9 1 2 1 2 1 1 2 1 2 1 9 4 3 12 1 38
398	19 1 18 1 38
399	19 1 38

Continued Fraction for square root of n for $2 < n < 500$ (non-square n)

401	20 40
402	20 20 40
403	20 13 2 1 3 1 3 1 2 13 40
404	20 10 40
405	20 8 40
406	20 6 1 2 4 7 1 4 1 7 4 2 1 6 40
407	20 5 1 2 1 5 40
408	20 5 40
409	20 4 2 7 1 1 1 4 2 2 13 13 2 2 4 1 1 1 7 2 4 40
410	20 4 40
411	20 3 1 1 1 19 1 1 1 3 40
412	20 3 2 1 3 1 4 3 2 13 10 13 2 3 4 1 3 1 2 3 40
413	20 3 9 1 4 1 9 3 40
414	20 2 1 7 2 7 1 2 40
415	20 2 1 2 4 6 1 1 3 1 1 6 4 2 1 2 40
416	20 2 1 1 9 1 1 2 40
417	20 2 2 1 1 1 5 4 1 12 1 4 5 1 1 1 2 2 40
418	20 2 4 20 4 2 40
419	20 2 7 1 2 3 1 2 1 19 1 2 1 3 2 1 7 2 40
420	20 2 40
421	20 1 1 13 5 1 3 1 2 1 1 1 2 9 1 7 3 3 2 2 3 3 7 1 9 2 1 1 1 2 1 3 1 5 13 1 1 40
422	20 1 1 5 2 1 3 20 3 1 2 5 1 1 40
423	20 1 1 3 4 3 1 1 40
424	20 1 1 2 4 5 1 1 1 9 1 1 1 5 4 2 1 1 40
425	20 1 1 1 1 1 1 40

Continued Fraction for square root of n for $2 < n < 500$ (non-square n)

n	Continued fraction
426	20 1 1 1 3 2 6 2 3 1 1 1 40
427	20 1 1 1 40
428	20 1 2 4 1 5 10 5 1 4 2 1 40
429	20 1 2 2 9 1 12 1 9 2 2 1 40
430	20 1 2 1 3 1 6 8 6 1 3 1 2 1 40
431	20 1 3 5 1 2 7 1 19 1 7 2 1 5 3 1 40
432	20 1 3 1 1 1 3 1 40
433	20 1 4 4 2 2 1 3 13 1 1 1 1 13 3 1 2 2 4 4 1 40
434	20 1 4 1 40
435	20 1 5 1 40
436	20 1 7 2 1 1 1 13 3 2 2 5 1 1 4 10 4 1 1 5 2 2 3 13 1 1 1 2 7 1 40
437	20 1 9 2 9 1 40
438	20 1 12 1 40
439	20 1 19 1 40
440	20 1 40
442	21 42
443	21 21 42
444	21 14 42
445	21 10 1 1 10 42
446	21 8 2 2 1 3 1 1 20 1 1 3 1 2 2 8 42
447	21 7 42
448	21 6 42
449	21 5 3 1 1 1 7 1 5 5 1 7 1 1 1 3 5 42
450	21 4 1 2 4 2 1 4 42

Continued Fraction for square root of n for $2 < n < 500$ (non-square n)

451	21 4 4 2 8 21 8 2 4 4 42
452	21 3 1 5 3 10 3 5 1 3 42
453	21 3 1 1 10 14 10 1 1 3 42
454	21 3 3 1 13 2 3 2 1 1 4 6 1 7 1 1 1 20 1 1 1 7 1 6 4 1 1 2 3 2 13 1 3 3 42
455	21 3 42
456	21 2 1 4 1 2 42
457	21 2 1 1 1 5 2 13 1 3 1 4 1 1 4 1 3 1 13 2 5 1 1 1 2 42
458	21 2 2 42
459	21 2 2 1 4 21 4 1 2 2 42
460	21 2 4 3 1 2 10 2 1 3 4 2 42
461	21 2 8 10 1 1 1 1 1 1 1 1 10 8 2 42
462	21 2 42
463	21 1 1 13 1 5 4 1 1 1 1 2 2 6 1 3 21 3 1 6 2 2 1 1 1 1 4 5 1 13 1 1 42
464	21 1 1 5 1 1 1 5 1 1 42
465	21 1 1 3 2 2 2 3 1 1 42
466	21 1 1 2 2 1 2 5 1 3 1 20 1 3 1 5 2 1 2 2 1 1 42
467	21 1 1 1 1 3 3 21 3 3 1 1 1 1 42
468	21 1 1 1 2 1 1 1 42
469	21 1 1 1 10 6 10 1 1 1 42
470	21 1 2 8 2 1 42
471	21 1 2 2 1 3 4 14 4 3 1 2 2 1 42
472	21 1 2 1 1 1 4 5 4 1 1 1 2 1 42
473	21 1 2 1 42
474	21 1 3 2 1 1 1 6 1 1 1 2 3 1 42
475	21 1 3 1 6 2 6 1 3 1 42

Continued Fraction for square root of n for $2 < n < 500$ (non-square n)

476	21 1 4 2 10 2 4 1 42
477	21 1 5 3 1 4 10 1 2 2 4 2 2 1 10 4 1 3 5 1 42
478	21 1 6 3 4 1 1 5 1 2 3 1 1 1 1 1 13 1 20 1 13 1 1 1 1 1 3 2 1 5 1 1 4 3 6 1 42
479	21 1 7 1 3 2 21 2 3 1 7 1 42
480	21 1 9 1 42
481	21 1 13 1 1 1 4 4 1 1 1 13 1 42
482	21 1 20 1 42
483	21 1 42
485	22 44
486	22 22 44
487	22 14 1 2 4 1 1 3 2 5 1 6 1 1 21 1 1 6 1 5 2 3 1 1 4 2 1 14 44
488	22 11 44
489	22 8 1 4 1 1 1 3 2 1 2 14 2 1 2 3 1 1 1 4 1 8 44
490	22 7 2 1 4 4 4 1 2 7 44
491	22 6 3 4 8 1 1 1 2 1 1 21 1 1 2 1 1 1 8 4 3 6 44
492	22 5 1 1 10 1 1 5 44
493	22 4 1 10 3 3 10 1 4 44
494	22 4 2 2 1 2 1 2 2 4 44
495	22 4 44
496	22 3 1 2 4 1 1 2 2 2 1 1 4 2 1 3 44
497	22 3 2 2 5 6 5 2 2 3 44
498	22 3 6 22 6 3 44
499	22 2 1 21 1 2 44
500	22 2 1 3 2 1 1 10 1 1 2 3 1 2 44

Appendix 4

"Least solution" of Pell's Equation $X^2 - NY^2 = 1$

N	X	Y
2	3	2
3	2	1
5	9	4
6	5	2
7	8	3
8	3	1
10	19	6
11	10	3
12	7	2
13	649	180
14	15	4
15	4	1
17	33	8
18	17	4
19	170	39
20	9	2
21	55	12
22	197	42
23	24	5
24	5	1

"Least solution" of Pell's Equation $X^2 - NY^2 = 1$

N	X	Y
26	51	10
27	26	5
28	127	24
29	9801	1820
30	11	2
31	1520	273
32	17	3
33	23	4
34	35	6
35	6	1
37	73	12
38	37	6
39	25	4
40	19	3
41	2049	320
42	13	2
43	3482	531
44	199	30
45	161	24
46	24335	3588
47	48	7
48	7	1
50	99	14

"Least solution" of Pell's Equation $X^2 - NY^2 = 1$

N	X	Y
51	50	7
52	649	90
53	66249	9100
54	485	66
55	89	12
56	15	2
57	151	20
58	19603	2574
59	530	69
60	31	4
61	1766319049	226153980
62	63	8
63	8	1
65	129	16
66	65	8
67	48842	5967
68	33	4
69	7775	936
70	251	30
71	3480	413
72	17	2
73	2281249	267000
74	3699	430
75	26	3

"Least solution" of Pell's Equation $X^2 - NY^2 = 1$

N	X	Y
76	57799	6630
77	351	40
78	53	6
79	80	9
80	9	1
82	163	18
83	82	9
84	55	6
85	285769	30996
86	10405	1122
87	28	3
88	197	21
89	500001	53000
90	19	2
91	1574	165
92	1151	120
93	12151	1260
94	2143295	221064
95	39	4
96	49	5
97	62809633	6377352
98	99	10
99	10	1

"Least solution" of Pell's Equation $X^2 - NY^2 = 1$

N	X	Y
101	201	20
102	101	10
103	227528	22419
104	51	5
105	41	4
106	32080051	3115890
107	962	93
108	1351	130
109	158070671986249	15140424455100
110	21	2
111	295	28
112	127	12
113	1204353	113296
114	1025	96
115	1126	105
116	9801	910
117	649	60
118	306917	28254
119	120	11
120	11	1
122	243	22
123	122	11
124	4620799	414960
125	930249	83204

"Least solution" of Pell's Equation $X^2 - NY^2 = 1$

N	X	Y
126	449	40
127	4730624	419775
128	577	51
129	16855	1484
130	6499	570
131	10610	927
132	23	2
133	2588599	224460
134	145925	12606
135	244	21
136	35	3
137	6083073	519712
138	47	4
139	77563250	6578829
140	71	6
141	95	8
142	143	12
143	12	1
145	289	24
146	145	12
147	97	8
148	73	6
149	25801741449	2113761020
150	49	4

"*Least solution*" of Pell's Equation $X^2 - NY^2 = 1$

N	X	Y
151	1728148040	140634693
152	37	3
153	2177	176
154	21295	1716
155	249	20
156	25	2
157	46698728731849	3726964292220
158	7743	616
159	1324	105
160	721	57
161	11775	928
162	19601	1540
163	64080026	5019135
164	2049	160
165	1079	84
166	1700902565	132015642
167	168	13
168	13	1
170	339	26
171	170	13
172	24248647	1848942
173	2499849	190060
174	1451	110
175	2024	153

"Least solution" of Pell's Equation $X^2 - NY^2 = 1$

N	X	Y
176	199	15
177	62423	4692
178	1601	120
179	4190210	313191
180	161	12
181	2469645423824185801	183567298683461940
182	27	2
183	487	36
184	24335	1794
185	9249	680
186	7501	550
187	1682	123
188	4607	336
189	55	4
190	52021	3774
191	8994000	650783
192	97	7
193	6224323426849	448036604040
194	195	14
195	14	1
197	393	28
198	197	14
199	16266196520	1153080099
200	99	7

"Least solution" of Pell's Equation $X^2 - NY^2 = 1$

N	X	Y
201	515095	36332
202	19731763	1388322
203	57	4
204	4999	350
205	39689	2772
206	59535	4148
207	1151	80
208	649	45
209	46551	3220
210	29	2
211	278354373650	19162705353
212	66249	4550
213	194399	13320
214	695359189925	47533775646
215	44	3
216	485	33
217	3844063	260952
218	126003	8534
219	74	5
220	89	6
221	1665	112
222	149	10
223	224	15
224	15	1

"*Least solution*" of Pell's Equation $X^2 - NY^2 = 1$

N	X	Y
226	451	30
227	226	15
228	151	10
229	5848201	386460
230	91	6
231	76	5
232	19603	1287
233	1072400673	70255304
234	5201	340
235	46	3
236	561799	36570
237	228151	14820
238	11663	756
239	6195120	400729
240	31	2
241	10085143557001249	649641205044600
242	19601	1260
243	70226	4505
244	1766319049	113076990
245	51841	3312
246	88805	5662
247	85292	5427
248	63	4
249	8553815	542076
250	39480499	2496966

"Least solution" of Pell's Equation $X^2 - NY^2 = 1$

N	X	Y
251	3674890	231957
252	127	8
253	3222617399	202604220
254	255	16
255	16	1
257	513	32
258	257	16
259	847225	52644
260	129	8
261	192119201	11891880
262	104980517	6485718
263	139128	8579
264	65	4
265	73738369	4529712
266	685	42
267	2402	147
268	4771081927	291440214
269	13449	820
270	5291	322
271	115974983600	7044978537
272	33	2
273	727	44
274	3959299	239190
275	199	12

"Least solution" of Pell's Equation $X^2 - NY^2 = 1$

N	X	Y
276	7775	468
277	159150073798980475849	9562401173878027020
278	2501	150
279	1520	91
280	251	15
281	2262200630049	134951575480
282	2351	140
283	138274082	8219541
284	24220799	1437240
285	2431	144
286	561835	33222
287	288	17
288	17	1
290	579	34
291	290	17
292	2281249	133500
293	12320649	719780
294	4801	280
295	2024999	117900
296	3699	215
297	48599	2820
298	335473872499	19433479650
299	415	24
300	1351	78

"*Least solution*" of Pell's Equation $X^2 - NY^2 = 1$

N	X	Y
301	5883392537695	339113108232
302	4276623	246092
303	2524	145
304	57799	3315
305	489	28
306	35	2
307	88529282	5052633
308	351	20
309	64202725495	3652365444
310	848719	48204
311	16883880	957397
312	53	3
313	32188120829134849	1819380158564160
314	392499	22150
315	71	4
316	12799	720
317	248678907849	13967198980
318	107	6
319	12901780	722361
320	161	9
321	215	12
322	323	18
323	18	1
325	649	36

"Least solution" of Pell's Equation $X^2 - NY^2 = 1$

N	X	Y
326	325	18
327	217	12
328	163	9
329	2376415	131016
330	109	6
331	2785589801443970	153109862634573
332	13447	738
333	73	4
334	63804373719695	3491219999244
335	604	33
336	55	3
337	2063810353129713793	112422913565764752
338	114243	6214
339	97970	5321
340	285769	15498
341	10626551	575460
342	37	2
343	130576328	7050459
344	10405	561
345	6761	364
346	17299	930
347	641602	34443
348	1567	84
349	169648201	9081060
350	449	24

"Least solution" of Pell's Equation $X^2 - NY^2 = 1$

N	X	Y
351	62425	3332
352	77617	4137
353	10157115393	540608704
354	258065	13716
355	954809	50676
356	500001	26500
357	3401	180
358	176579805797	9332532726
359	360	19
360	19	1
362	723	38
363	362	19
364	4954951	259710
365	23915529	1251796
366	907925	47458
367	19019995568	992835687
368	1151	60
369	8396801	437120
370	213859	11118
371	1695	88
372	12151	630
373	52387849	2712540
374	3365	174
375	15124	781

"Least solution" of Pell's Equation $X^2 - NY^2 = 1$

N	X	Y
376	2143295	110532
377	233	12
378	8749	450
379	12941197220540690	664744650125541
380	39	2
381	1015	52
382	164998439999	8442054600
383	18768	959
384	4801	245
385	95831	4884
386	111555	5678
387	3482	177
388	62809633	3188676
389	3287049	166660
390	79	4
391	7338680	371133
392	99	5
393	46437143	2342444
394	312086396361222451	15722685507826110
395	159	8
396	199	10
397	838721786045180184649	42094239791738433660
398	399	20
399	20	1

"*Least solution*" of Pell's Equation $X^2 - NY^2 = 1$

N	X	Y
401	801	40
402	401	20
403	669878	33369
404	201	10
405	161	8
406	59468095	2951352
407	2663	132
408	101	5
409	25052977273092427986049	1238789998647218582160
410	81	4
411	49730	2453
412	103537981567	5100950232
413	113399	5580
414	24335	1196
415	18412804	903849
416	5201	255
417	85322647	4178268
418	33857	1656
419	270174970	13198911
420	41	2
421	3879474045914926879468217167061449	189073995951839020880499780706260
422	7022501	341850
423	4607	224
424	32080051	1557945
425	143649	6968

"Least solution" of Pell's Equation $X^2 - NY^2 = 1$

N	X	Y
426	88751	4300
427	62	3
428	1850887	89466
429	1524095	73584
430	2862251	138030
431	151560720	7300423
432	1351	65
433	104564907854286695713	5025068784834899736
434	125	6
435	146	7
436	158070671986249	7570212227550
437	4599	220
438	293	14
439	440	21
440	21	1
442	883	42
443	442	21
444	295	14
445	43468489	2060604
446	110166015	5216512
447	148	7
448	127	6
449	71798771299708449	3388393513402120
450	19601	924

"*Least solution*" of Pell's Equation $X^2 - NY^2 = 1$

N	X	Y
451	46471490	2188257
452	1204353	56648
453	1653751	77700
454	16916040084175685	793909098494766
455	64	3
456	1025	48
457	6983244756398928218113	326662411570389853632
458	22899	1070
459	499850	23331
460	2535751	118230
461	1182351890184201	55067617520620
462	43	2
463	247512720456368	11502891625161
464	9801	455
465	15871	736
466	938319425	43466808
467	1625626	75225
468	649	30
469	137215	6336
470	1691	78
471	7838695	361188
472	306917	14127
473	87	4
474	193549	8890
475	57799	2652

"*Least solution*" of Pell's Equation $X^2 - NY^2 = 1$

N	X	Y
476	28799	1320
477	8777860001	401910600
478	1617319577991743	73974475657896
479	2989440	136591
480	241	11
481	1859131879201	84769117080
482	483	22
483	22	1
485	969	44
486	485	22
487	51906073840568	2352088722477
488	243	11
489	7592629975	343350596
490	1039681	46968
491	93628044170	4225374483
492	29767	1342
493	935662752649	42140131020
494	73035	3286
495	89	4
496	4620799	207480
497	1201887	53912
498	179777	8056
499	4490	201
500	930249	41602

Appendix 5

Test of Divisibility by 7, 11, and 13.

- Divisibility test for 7.

Let N be a any large natural number. If the number of its digits is not a multiple of 3 we can make it so by adding one or two zeros on the left. So we may assume that N is a $3n$-digit number. We think of its $3n$ digits as consisting of n blocks each of size three. Let $a_{i,j}$, where $1 \leq i \leq n$ and $0 \leq j \leq 2$ be the $3n$ digits of N counting from right to left. That is

$$N = a_{n,2}a_{n,1}a_{n,0} \cdots \cdots a_{i,2}a_{i,1}a_{i,0} \cdots \cdots a_{1,2}a_{1,1}a_{1,0}$$

Since $10 = 7 + 3 \equiv 3 \pmod 7$. Also $10^3 \equiv 3^3 \equiv -1 \pmod 7$. We will find value of N modulo 7 as follows

$$\begin{aligned}
N &= \textstyle\sum_{i=1}^{i=n} \sum_{j=0}^{j=2} a_{i,j} 10^{3(i-1)+j} \\
&= \textstyle\sum_{i=1}^{i=n} \sum_{j=0}^{j=2} a_{i,j} 10^{3(i-1)} \cdot 10^j \\
&= \textstyle\sum_{i=1}^{i=n} \sum_{j=0}^{j=2} a_{i,j} 3^{3(i-1)} \cdot 3^j \pmod 7 \\
&= \textstyle\sum_{i=1}^{i=n} \sum_{j=0}^{j=2} a_{i,j} 3^{3^{(i-1)}} \cdot 3^j \pmod 7 \\
&= \textstyle\sum_{i=1}^{i=n} \sum_{j=0}^{j=2} a_{i,j} (3^3)^{(i-1)} \cdot 3^j \pmod 7 \\
&= \textstyle\sum_{i=1}^{i=n} \sum_{j=0}^{j=2} a_{i,j} (-1)^{(i-1)} \cdot 3^j \pmod 7 \\
&= \textstyle\sum_{i=1}^{i=n} \left(a_{i,0}(-1)^{(i-1)} + a_{i,1}(-1)^{i-1} \cdot 3 + a_{i,2}(-1)^{i-1} \cdot 2\right) \pmod 7 \\
&= \textstyle\sum_{i=1}^{i=n} (-1)^{i-1}(a_{i,0} + 3 \cdot a_{i,1} + 2 \cdot a_{i,2}) \pmod 7 \\
&= \textstyle\sum_{i=1}^{i=n} (-1)^{i-1}(\mathbf{1} \cdot a_{i,0} + \mathbf{3} \cdot a_{i,1} + \mathbf{2} \cdot a_{i,2}) \pmod 7.
\end{aligned}$$

The above description of the given number modulo 7 suggests the following method of calculating N modulo 7.

Partition the $3n$ digits of N into n blocks of size three each. Number the blocks 1 to n from *right* to *left*. On the top of the numbers of each block write 1, 3, and 2 from *right to left*. For each block multiply its content by the number on the top and add. If you wish you may reduce the resulting number modulo 7 - that is, divide it by seven and replace it by the remainder. Let N_1 be the sum of the contents of the odd-numbered blocks and N_2 is the sum of the contents of the even-numbered blocks then $N \equiv N_1 - N_2 \pmod 7$.

Example 1

Let us find the remainder when the 12-digit number $N = 123456789241$ is divided by 7. Since N has 12 digits which happens to be a multiple of 3 we do not have to pad it on the left with zeros. We write the number N as consisting of four blocks, each of size three, and on the top of the digits of each block we write 1, 3, and 2 *from right to left* and multiply the digits in each column and write the result underneath (that is in the next row) and add the contents of the three numbers in each block in the last row. We have

2	**3**	**1**	**2**	**3**	**1**	**2**	**3**	**1**	**2**	**3**	**1**
1	2	3	4	5	6	7	8	9	2	4	1
2	6	3	8	15	6	14	24	9	4	12	1
	11			29			47			17	

N_1 = Sum of first and third blocks (from right to left) $= 17 + 29 = 46$.

N_2 = Sum of second and fourth blocks (from right to left) $= 47 + 11 = 58$.

As $N_1 - N_2 = 46 - 58 = -12 \equiv 2 \pmod 7$, the number N leaves a remainder of 2 when divided by 7.

Example 2

Let us find the remainder when the 10-digit number $N = 1234759680$ is divided by 7. Since this number has 10 digits we pad this number on the left by two zeros so the number reads $N = 001234759680$. Proceeding as before we have

2	**3**	**1**	**2**	**3**	**1**	**2**	**3**	**1**	**2**	**3**	**1**
0	0	1	2	3	4	7	5	9	6	8	0
0	0	1	4	9	4	14	15	9	12	24	0
	1			17			38			36	

N_1 = Sum of first and third blocks (from right to left) $= 36 + 17 = 53$.

N_2 = Sum of second and fourth blocks (from right to left) $= 38 + 1 = 39$.

As $N_1 - N_2 = 53 - 39 = 14$, so the number N is divisible by 7.

Note: It is an easy exercise to show that $10x + y \equiv 0 \mod 7 \iff x - 2y \equiv 0 \pmod 7$. This observation can also be used to test if a given number is divisible by 7 or not. However, this method will not give the remainder when N is not divisible by 7.

- **Divisibility test for** 11.

Let $N = a_{2n}a_{2n-1}\cdots a_2a_1$ be a $2n$-digit number. If the given number has odd number of digits to begin with we may assume that $a_{2n} = 0$. We have

$$\begin{aligned} N &= \sum_{i=1}^{i=2n} a_i 10^{i-1} \\ &= \sum_{i=1}^{i=2n} a_i (11-1)^{i-1} \\ &\equiv \sum_{i=1}^{i=2n} a_i (-1)^{i-1} \pmod{11} \\ &\equiv (a_1 + a_3 + \cdots + a_{2n-1}) - (a_2 + a_4 + \cdots + a_{2n}) \pmod{11} \end{aligned}$$

This shows that the number N is divisible by 11 if and only if the number obtained by taking the difference of the sums of the alternate digits of N is divisible by 11.

- Divisibility test for 13.

Let N be a any large natural. If the number of its digits is not a multiple of 3 we can make it so by adding one or two zeros on the left. So we may assume that N is a $3n$-digit number. We think of its $3n$ digits as consisting of n blocks each of size three. Let $a_{i,j}$, where $1 \le i \le n$ and $0 \le j \le 2$ be the $3n$ digits of N counting from right to left. That is

$$N = a_{n,2}a_{n,1}a_{n,0}\cdots\cdots a_{i,2}a_{i,1}a_{i,0}\cdots\cdots a_{1,2}a_{1,1}a_{1,0}$$

Since $10 = 13 - 3 \equiv -3 \pmod{13}$, $10^3 \equiv (-3)^3 \equiv -1 \pmod{13}$. We will find value of N modulo 7 as follows

$$\begin{aligned} N &= \sum_{i=1}^{i=n}\sum_{j=0}^{j=2} a_{i,j} 10^{3(i-1)+j} \\ &= \sum_{i=1}^{i=n}\sum_{j=0}^{j=2} a_{i,j} 10^{3(i-1)} \cdot 10^j \\ &= \sum_{i=1}^{i=n}\sum_{j=0}^{j=2} a_{i,j} (-3)^{3(i-1)} \cdot (-3)^j \pmod{13} \\ &= \sum_{i=1}^{i=n}\sum_{j=0}^{j=2} a_{i,j} (-1)^{(i-1)} \cdot (-3)^j \pmod{13} \\ &= \sum_{i=1}^{i=n} \left(a_{i,0}(-1)^{(i-1)} + a_{i,1}(-1)^{i-1}\cdot(-3) + a_{i,2}(-1)^{i-1}\cdot 9\right) \pmod{13} \\ &= \sum_{i=1}^{i=n}(-1)^{i-1}(a_{i,0} + 10\cdot a_{i,1} + 9\cdot a_{i,2}) \pmod{13} \\ &= \sum_{i=1}^{i=n}(-1)^{i-1}(\mathbf{1}\cdot a_{i,0} + \mathbf{10}\cdot a_{i,1} + \mathbf{9}\cdot a_{i,2}) \pmod{13}. \end{aligned}$$

The above description of the given number modulo 13 suggests the following method of calculating N modulo 13.

Partition the $3n$ digits of N into n blocks of size three each. Number the blocks 1 to n from *right* to *left*. On the top of the numbers of each block write 1, 10, and 9 from *right to left*. For each block multiply its contents by the number on the top and add. If you wish you may reduce the resulting number modulo 13 - that is, divide it by 13 and replace it by the remainder. Let N_1 be the sum of the contents of the odd-numbered blocks and N_2 is the sum of contents of the even-numbered blocks then $N \equiv N_1 - N_2 \pmod{13}$.

Example 1

Let us find the remainder when the 12-digit number $N = 123456789241$ is divided by 13. Since N has 12 digits which happens to be a multiple of 3 we do not have to pad it on the left with zeros. We write the number N as consisting of four blocks, each of size three, and on the top of the digits of each block we write 1, 10 and 9 *from right to left* and multiply the digits in each column and write the result underneath (that is in the next row) and add the contents of the three numbers in each block in the last row. We have

9	**10**	**1**	**9**	**10**	**1**	**9**	**10**	**1**	**9**	**10**	**1**
1	2	3	4	5	6	7	8	9	2	4	1
9	20	3	36	50	6	63	80	9	18	40	1
	32			92			152			59	

N_1 = Sum of first and third blocks (from right to left) $= 59 + 92 = 151$.
N_2 = Sum of second and fourth blocks (from right to left) $= 32 + 152 = 182$.
As $N_1 - N_2 = 151 - 184 = -33 \equiv 6 \pmod{13}$, the number N leaves a remainder of 6 when divided by 13.

Example 2

Let us find the remainder when the 10-digit number $N = 1234759680$ is divided by 13. Since this number has 10 digits we pad this number on the left by two zero so the number reads $N = 001234759680$. In this case we proceed slightly differently. Since $9 \equiv -4 \mod 13$ and $10 \equiv -3 \mod 13$, instead of using $1, 10$, and 9 as block multipliers we use the equivalent $1, -3$, and -4. We also reduce the contents of the last row modulo 13. Proceeding as before we have

−4	**−3**	**1**	**−4**	**−3**	**1**	**−4**	**−3**	**1**	**−4**	**−3**	**1**
0	0	1	2	3	4	7	5	9	6	8	0
0	0	1	−8	−9	4	−28	−15	9	−24	−24	0
	1			−13			−34			−48	
	1			0			5			4	

N_1 = Sum of first and third blocks (from right to left) $= 4 + 0 = 4$.

N_2 = Sum of second and fourth blocks (from right to left) $= 5 + 1 = 6$.

As $N_1 - N_2 = 4 - 6 = -2 \equiv 11 \mod 13$, so the remainder when N is divided by 13 is 11.

Note: It is an easy exercise to show that $10x + y \equiv 0 \mod 13 \iff x + 4y \equiv 0 \pmod{13}$. This observation can be used to test if a given number is divisible by 13 or not. However, this method will not give the remainder when N is not divisible by 13.

Appendix 6

Calculating Square Roots

We first observe that the integer part of the square root of a $2n$-digit number is a n digit number. Let $N = a_1b_1a_2b_2\cdots a_ib_i\cdots a_nb_n$ be a $2n$-digit number. Here we allow a_1 to be zero so that any number with an odd number of digits can be padded with one zero as a leading digit to make it an even digit number. Suppose that $[\sqrt{N}] = x_1x_2\cdots x_ix_{i+1}\cdots x_n$, where $x_i \in \{0,1,2,\cdots,9\}$ for $1 \le i \le n$, but $x_1 \ne 0$ and Clearly $x_1 = [\sqrt{10a_1+b_1}]$.
Suppose that we have already calculated the digits $x_1, x_2, \cdots x_i$ for $i < n$. We shall show how to calculate the digit x_{i+1}. Let

$$\mathbf{X} = x_1x_2\cdots x_i = x_1 10^{i-1} + x_2 10^{i-2} + \cdots + x_i.$$

Then

$$\left(\mathbf{X}\cdot 10^{n-i}\right)^2 \le N < \left((\mathbf{X}+\mathbf{1})\cdot 10^{n-i}\right)^2.$$

Then the next digit $y = x_{i+1}$ is the largest digit satisfying

$$\left[(10\mathbf{X}+y)\cdot 10^{n-i-1}\right]^2 \le N, \text{ or } \left(100\mathbf{X^2} + 20\mathbf{X}y + y^2\right)\cdot 10^{2n-2i-2} \le N$$

or

$$\mathbf{X^2}\cdot 10^{2n-2i} + (20\mathbf{X}+y)y\cdot 10^{2n-2i-2} \le N \text{ or } (20\mathbf{X}+y)y\cdot 10^{2n-2i-2} \le N - \mathbf{X^2}\cdot 10^{2n-2i}$$

That is, the digit $y = x_{i+1}$ is the largest integer in the set $\{0,1,2,3,4,5,6,7,8,9\}$ satisfying the condition

$$(20\mathbf{X}+y)y \le \frac{N}{10^{2n-2i-2}} - \mathbf{X^2}\cdot 10^2$$

Let us calculate the square root of the number 123456 correct up to three decimal places using the above procedure. We let $N = 123456000000$ and will calculate the integer part of the square root of N and then divide it by 10^3 to get the square root of 123456 correct up to three decimal places. The following table illustrates the various steps of the algorithm. As our $N = 123456000000$ has 12 digits $n = 6$.

i	x_i	$\mathbf{X}$	$\left[\frac{N}{10^{2n-2i-2}}\right] - \mathbf{X^2} \cdot 10^2 = \left[\frac{123456}{10^{6-2i}}\right] - 100\mathbf{X^2}$	$y = x_{i+1}$	$20\mathbf{X} + y$
1	3	3	$1234 - 900 = 334(65 \times 5 = 325)$	5	65
2	5	35	$123456 - 122500 = 956(701 \times 1 = 701)$	1	701
3	1	351	$12345600 - 12320100 = 25500(7023 \times 3 = 21069)$	3	7023
4	3	3513	$1234560000 - 1234116900 = 443100(70266 \times 6 = 421596)$	6	70266
5	6	35136	$123456000000 - 123453849600 = 2150400(*)$	3	702723
6	3	351363			

$(*)702723 \times 3 = 2108169.$

Therefore $\sqrt{123456} \approx 351.363$

Appendix 7

List of primes below 10,000

How to Read the Prime Number Cards

1) Every black dot appears in one of 5 locations in a square.

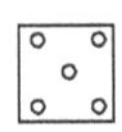

2) Choose a black dot.

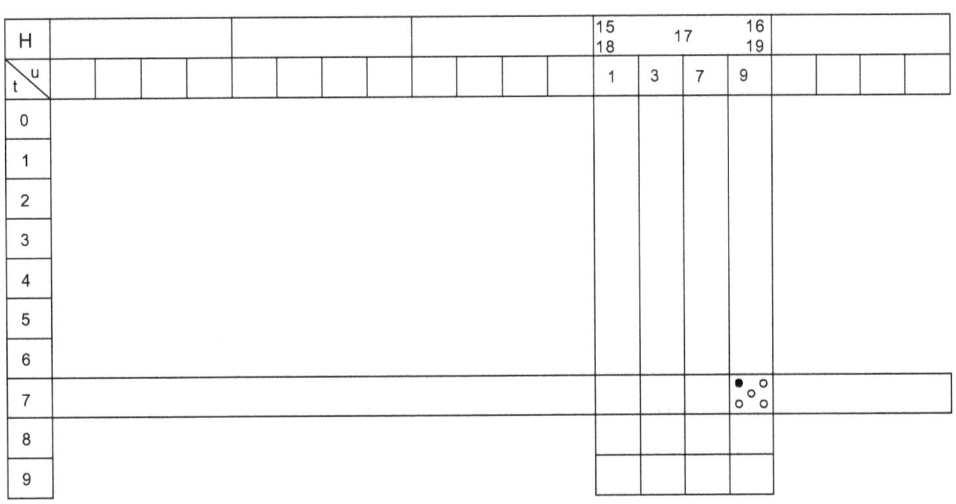

3) Look at the very top of the card. Since it's in the upper left hand corner, it corresponds to 15 hundreds.

4) The number of units is in the second row above the dot. So it's 9.

5) The number of tens is to the far left of the dot and in the same row as the dot. So it's 7.

6) The prime number is 1579.

$0 < p < 2{,}500$

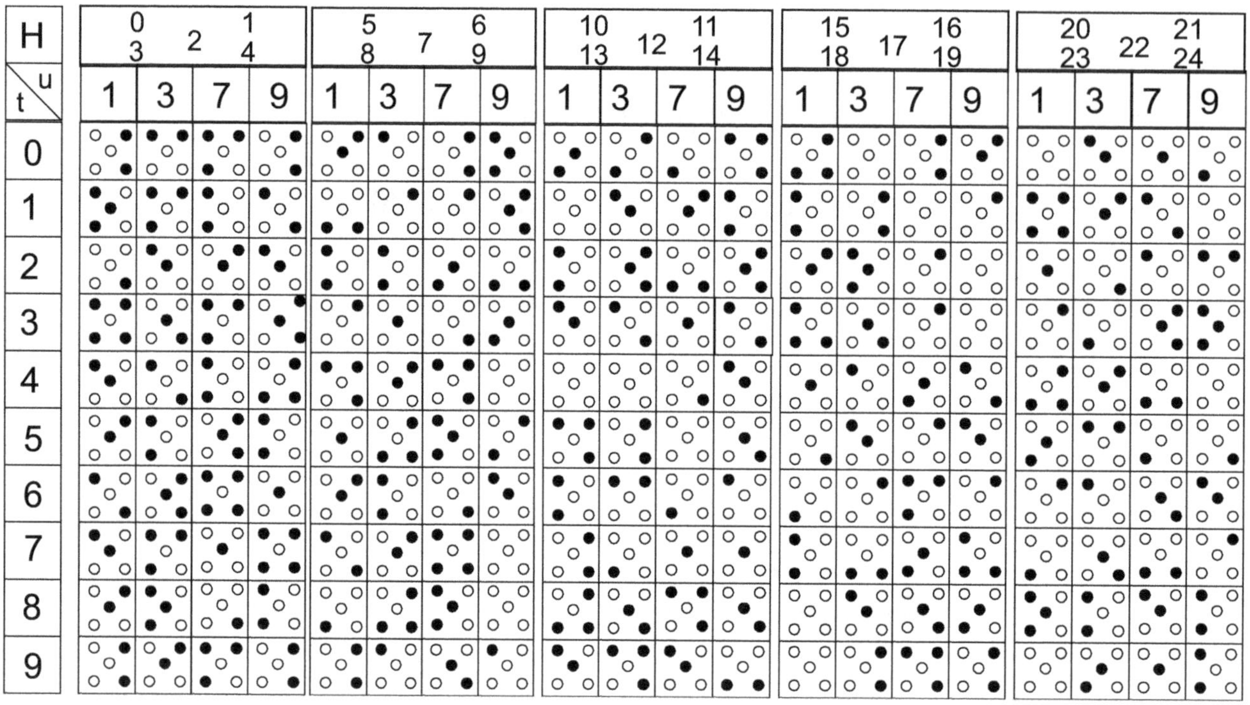

$2500 < p < 5000$

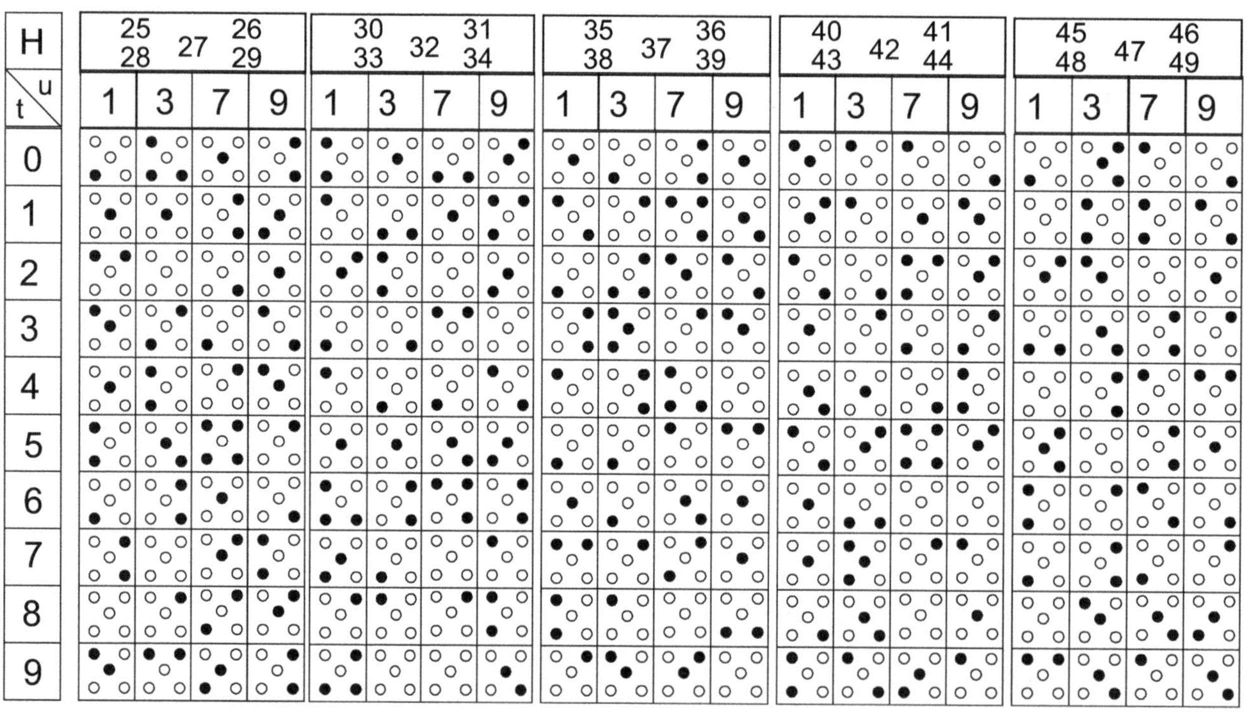

5000 < p < 7500

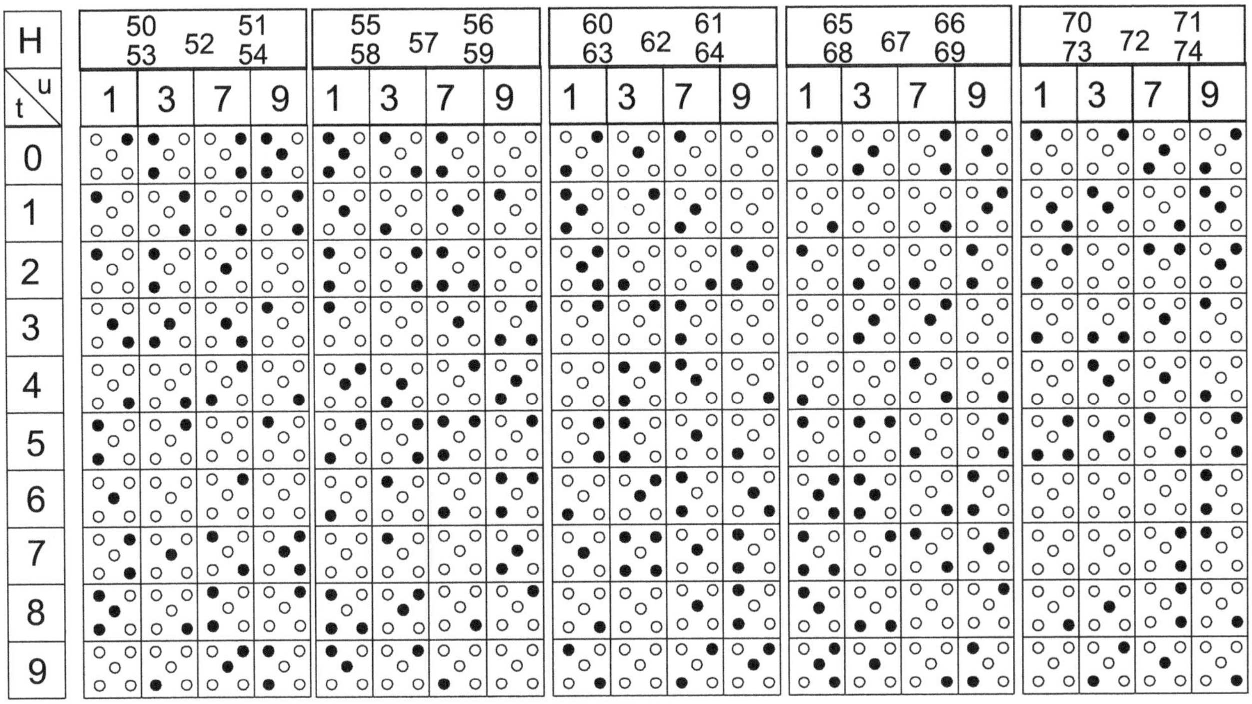

7500 < p < 10000

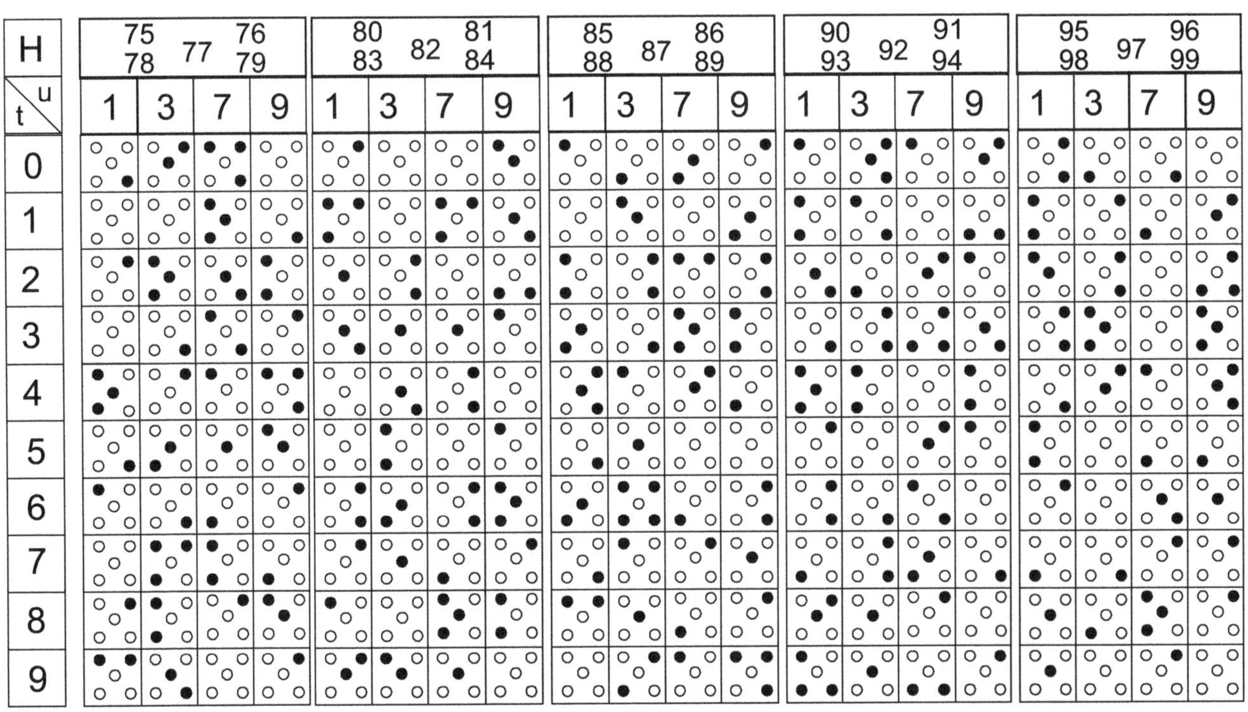

APPENDIX 7

Prime Numbers up to 10,000

2, 3, 5, 7, 11, 13, 17, 19, 23, 29, 31, 37, 41, 43, 47, 53, 59, 61, 67, 71, 73, 79, 83, 89, 97, 101, 103, 107, 109, 113, 127, 131, 137, 139, 149, 151, 157, 163, 167, 173, 179, 181, 191, 193, 197, 199, 211, 223, 227, 229, 233, 239, 241, 251, 257, 263, 269, 271, 277, 281, 283, 293, 307, 311, 313, 317, 331, 337, 347, 349, 353, 359, 367, 373, 379, 383, 389, 397, 401, 409, 419, 421, 431, 433, 439, 443, 449, 457, 461, 463, 467, 479, 487, 491, 499, 503, 509, 521, 523, 541, 547, 557, 563, 569, 571, 577, 587, 593, 599, 601, 607, 613, 617, 619, 631, 641, 643, 647, 653, 659, 661, 673, 677, 683, 691, 701, 709, 719, 727, 733, 739, 743, 751, 757, 761, 769, 773, 787, 797, 809, 811, 821, 823, 827, 829, 839, 853, 857, 859, 863, 877, 881, 883, 887, 907, 911, 919, 929, 937, 941, 947, 953, 967, 971, 977, 983, 991, 997, 1009, 1013, 1019, 1021, 1031, 1033, 1039, 1049, 1051, 1061, 1063, 1069, 1087, 1091, 1093, 1097, 1103, 1109, 1117, 1123, 1129, 1151, 1153, 1163, 1171, 1181, 1187, 1193, 1201, 1213, 1217, 1223, 1229, 1231, 1237, 1249, 1259, 1277, 1279, 1283, 1289, 1291, 1297, 1301, 1303, 1307, 1319, 1321, 1327, 1361, 1367, 1373, 1381, 1399, 1409, 1423, 1427, 1429, 1433, 1439, 1447, 1451, 1453, 1459, 1471, 1481, 1483, 1487, 1489, 1493, 1499, 1511, 1523, 1531, 1543, 1549, 1553, 1559, 1567, 1571, 1579, 1583, 1597, 1601, 1607, 1609, 1613, 1619, 1621, 1627, 1637, 1657, 1663, 1667, 1669, 1693, 1697, 1699, 1709, 1721, 1723, 1733, 1741, 1747, 1753, 1759, 1777, 1783, 1787, 1789, 1801, 1811, 1823, 1831, 1847, 1861, 1867, 1871, 1873, 1877, 1879, 1889, 1901, 1907, 1913, 1931, 1933, 1949, 1951, 1973, 1979, 1987, 1993, 1997, 1999, 2003, 2011, 2017, 2027, 2029, 2039, 2053, 2063, 2069, 2081, 2083, 2087, 2089, 2099, 2111, 2113, 2129, 2131, 2137, 2141, 2143, 2153, 2161, 2179, 2203, 2207, 2213, 2221, 2237, 2239, 2243, 2251, 2267, 2269, 2273, 2281, 2287, 2293, 2297, 2309, 2311, 2333, 2339, 2341, 2347, 2351, 2357, 2371, 2377, 2381, 2383, 2389, 2393, 2399, 2411, 2417, 2423, 2437, 2441, 2447, 2459, 2467, 2473, 2477, 2503, 2521, 2531, 2539, 2543, 2549, 2551, 2557, 2579, 2591, 2593, 2609, 2617, 2621, 2633, 2647, 2657, 2659, 2663, 2671, 2677, 2683, 2687, 2689, 2693, 2699, 2707, 2711, 2713, 2719, 2729, 2731, 2741, 2749, 2753, 2767, 2777, 2789, 2791, 2797, 2801, 2803, 2819, 2833, 2837, 2843, 2851, 2857, 2861, 2879, 2887, 2897, 2903, 2909, 2917, 2927, 2939, 2953, 2957, 2963, 2969, 2971, 2999, 3001, 3011, 3019, 3023, 3037, 3041, 3049, 3061, 3067, 3079, 3083, 3089, 3109, 3119, 3121, 3137, 3163, 3167, 3169, 3181, 3187, 3191, 3203, 3209, 3217, 3221, 3229, 3251, 3253, 3257, 3259, 3271, 3299, 3301, 3307, 3313, 3319, 3323, 3329, 3331, 3343, 3347, 3359, 3361, 3371, 3373, 3389, 3391, 3407, 3413, 3433, 3449, 3457, 3461, 3463, 3467, 3469, 3491, 3499, 3511, 3517, 3527, 3529, 3533, 3539, 3541, 3547, 3557, 3559, 3571, 3581, 3583, 3593, 3607, 3613, 3617, 3623, 3631, 3637, 3643, 3659, 3671, 3673, 3677, 3691, 3697, 3701, 3709, 3719, 3727, 3733, 3739, 3761, 3767, 3769, 3779, 3793, 3797, 3803, 3821, 3823, 3833, 3847, 3851, 3853, 3863, 3877, 3881, 3889, 3907, 3911, 3917, 3919, 3923, 3929, 3931, 3943, 3947, 3967, 3989, 4001, 4003, 4007, 4013, 4019, 4021, 4027, 4049, 4051, 4057, 4073, 4079, 4091, 4093, 4099, 4111, 4127, 4129, 4133, 4139, 4153, 4157, 4159, 4177, 4201, 4211, 4217, 4219, 4229, 4231, 4241, 4243, 4253, 4259, 4261, 4271, 4273, 4283, 4289, 4297, 4327, 4337, 4339, 4349, 4357, 4363, 4373, 4391, 4397, 4409, 4421, 4423, 4441, 4447, 4451, 4457, 4463, 4481, 4483, 4493, 4507, 4513, 4517, 4519, 4523, 4547, 4549, 4561, 4567, 4583, 4591, 4597, 4603, 4621, 4637, 4639, 4643, 4649, 4651, 4657, 4663, 4673, 4679, 4691, 4703, 4721, 4723, 4729, 4733, 4751, 4759, 4783, 4787, 4789, 4793, 4799, 4801, 4813, 4817, 4831, 4861, 4871, 4877, 4889, 4903, 4909, 4919, 4931, 4933, 4937, 4943, 4951, 4957, 4967, 4969, 4973, 4987, 4993,

4999, 5003, 5009, 5011, 5021, 5023, 5039, 5051, 5059, 5077, 5081, 5087, 5099, 5101, 5107, 5113, 5119, 5147, 5153, 5167, 5171, 5179, 5189, 5197, 5209, 5227, 5231, 5233, 5237, 5261, 5273, 5279, 5281, 5297, 5303, 5309, 5323, 5333, 5347, 5351, 5381, 5387, 5393, 5399, 5407, 5413, 5417, 5419, 5431, 5437, 5441, 5443, 5449, 5471, 5477, 5479, 5483, 5501, 5503, 5507, 5519, 5521, 5527, 5531, 5557, 5563, 5569, 5573, 5581, 5591, 5623, 5639, 5641, 5647, 5651, 5653, 5657, 5659, 5669, 5683, 5689, 5693, 5701, 5711, 5717, 5737, 5741, 5743, 5749, 5779, 5783, 5791, 5801, 5807, 5813, 5821, 5827, 5839, 5843, 5849, 5851, 5857, 5861, 5867, 5869, 5879, 5881, 5897, 5903, 5923, 5927, 5939, 5953, 5981, 5987, 6007, 6011, 6029, 6037, 6043, 6047, 6053, 6067, 6073, 6079, 6089, 6091, 6101, 6113, 6121, 6131, 6133, 6143, 6151, 6163, 6173, 6197, 6199, 6203, 6211, 6217, 6221, 6229, 6247, 6257, 6263, 6269, 6271, 6277, 6287, 6299, 6301, 6311, 6317, 6323, 6329, 6337, 6343, 6353, 6359, 6361, 6367, 6373, 6379, 6389, 6397, 6421, 6427, 6449, 6451, 6469, 6473, 6481, 6491, 6521, 6529, 6547, 6551, 6553, 6563, 6569, 6571, 6577, 6581, 6599, 6607, 6619, 6637, 6653, 6659, 6661, 6673, 6679, 6689, 6691, 6701, 6703, 6709, 6719, 6733, 6737, 6761, 6763, 6779, 6781, 6791, 6793, 6803, 6823, 6827, 6829, 6833, 6841, 6857, 6863, 6869, 6871, 6883, 6899, 6907, 6911, 6917, 6947, 6949, 6959, 6961, 6967, 6971, 6977, 6983, 6991, 6997, 7001, 7013, 7019, 7027, 7039, 7043, 7057, 7069, 7079, 7103, 7109, 7121, 7127, 7129, 7151, 7159, 7177, 7187, 7193, 7207, 7211, 7213, 7219, 7229, 7237, 7243, 7247, 7253, 7283, 7297, 7307, 7309, 7321, 7331, 7333, 7349, 7351, 7369, 7393, 7411, 7417, 7433, 7451, 7457, 7459, 7477, 7481, 7487, 7489, 7499, 7507, 7517, 7523, 7529, 7537, 7541, 7547, 7549, 7559, 7561, 7573, 7577, 7583, 7589, 7591, 7603, 7607, 7621, 7639, 7643, 7649, 7669, 7673, 7681, 7687, 7691, 7699, 7703, 7717, 7723, 7727, 7741, 7753, 7757, 7759, 7789, 7793, 7817, 7823, 7829, 7841, 7853, 7867, 7873, 7877, 7879, 7883, 7901, 7907, 7919, 7927, 7933, 7937, 7949, 7951, 7963, 7993, 8009, 8011, 8017, 8039, 8053, 8059, 8069, 8081, 8087, 8089, 8093, 8101, 8111, 8117, 8123, 8147, 8161, 8167, 8171, 8179, 8191, 8209, 8219, 8221, 8231, 8233, 8237, 8243, 8263, 8269, 8273, 8287, 8291, 8293, 8297, 8311, 8317, 8329, 8353, 8363, 8369, 8377, 8387, 8389, 8419, 8423, 8429, 8431, 8443, 8447, 8461, 8467, 8501, 8513, 8521, 8527, 8537, 8539, 8543, 8563, 8573, 8581, 8597, 8599, 8609, 8623, 8627, 8629, 8641, 8647, 8663, 8669, 8677, 8681, 8689, 8693, 8699, 8707, 8713, 8719, 8731, 8737, 8741, 8747, 8753, 8761, 8779, 8783, 8803, 8807, 8819, 8821, 8831, 8837, 8839, 8849, 8861, 8863, 8867, 8887, 8893, 8923, 8929, 8933, 8941, 8951, 8963, 8969, 8971, 8999, 9001, 9007, 9011, 9013, 9029, 9041, 9043, 9049, 9059, 9067, 9091, 9103, 9109, 9127, 9133, 9137, 9151, 9157, 9161, 9173, 9181, 9187, 9199, 9203, 9209, 9221, 9227, 9239, 9241, 9257, 9277, 9281, 9283, 9293, 9311, 9319, 9323, 9337, 9341, 9343, 9349, 9371, 9377, 9391, 9397, 9403, 9413, 9419, 9421, 9431, 9433, 9437, 9439, 9461, 9463, 9467, 9473, 9479, 9491, 9497, 9511, 9521, 9533, 9539, 9547, 9551, 9587, 9601, 9613, 9619, 9623, 9629, 9631, 9643, 9649, 9661, 9677, 9679, 9689, 9697, 9719, 9721, 9733, 9739, 9743, 9749, 9767, 9769, 9781, 9787, 9791, 9803, 9811, 9817, 9829, 9833, 9839, 9851, 9857, 9859, 9871, 9883, 9887, 9901, 9907, 9923, 9929, 9931, 9941, 9949, 9967, 9973

Appendix 8

Finding Day of the Week
John Conway's Doomsday Algorithm

In this appendix we shall discuss John Conway's Doomsday Algorithm (1973) for finding the day of the week given any month, day, and year and provide several examples of the same. Let us agree to represent the days of the week from Sunday to Saturday by their digital equivalents 0 through 6 as they do with airline schedules. For describing John Conway's Doomsday Algorithm we shall take as an example of finding the day of the week for December 18, 1993. We write the given (four digit) year in the form CY where C is hundreds in the year and Y is the remaining part of the year. So for the year 1993 we have $C = 19$ and $Y = 93$. For any given date in a year, we define DMc (date and month contribution), Cc (century contribution), and Yc (year contribution) as follows:

- **Calculating** DMc

  Let us agree to define March 0 as the day before March 1 in any year. So for a leap year March 0 is same as February 29 and in a non-leap year it is same as February 28. It can be seen that in any year March 0, April 4, May 9, June 6, July 11, August 8, September 5, October 10, November 7, and December 12 fall on the same day of the week. This day has been referred to as Doomsday for the corresponding year by John Conway and we shall refer to the corresponding dates in various months as Conway numbers. As an aid for memorizing these numbers note that in even numbered months, excepting the month of February, these numbers are the same (even numbers) as the month number. For the rest the days in May and September correspond to each others number. That is the Conway number for the fifth month is nine and for the ninth month is five. Likewise for July and November. For the month of January and February the Conway numbers are 4 and 1 (respectively) for leap year and 3 and 0 (respectively) for non-leap year.

The following table gives the list of Conway Numbers for leap and non-leap years. Note that the Conway numbers for the leap and non-leap years differ only for the months of January and February. We can always change these mod 7 if we wish.

Month	Conway Number (Leap Year)	Conway Number (Non-leap Year)
January	4	3
February	1	0
March	0	0
April	4	4
May	9	9
June	6	6
July	11	11
August	8	8
September	5	5
October	10	10
November	7	7
December	12	12

Let d represent the day of the month in our problem. So for December 18, we have $d = 18$. For any date we define its DMc as $DMc = d -$ **Conway Number** for the corresponding month. For example for December 18 we have $DMc = 18 - 12 = \mathbf{6}$ and for September 2 we have $DMc = 2 - 5 = \mathbf{-3}$. The following table gives the value of DMc as a function of d.

	DMc (Leap Year)	DMc (Non-leap Year)
January	$d-4$	$d-3$
February	$d-1$	d
March	d	same
April	$d-4$	same
May	$d-9$	same
June	$d-6$	same
July	$d-11$	same
August	$d-8$	same
September	$d-5$	same
October	$d-10$	same
November	$d-7$	same
December	$d-12$	same

- **Calculating** Cc (Century Contribution - *Anchor Day or Number*)

  Suppose that the year lies in n-th century, that is $n = C + 1$. Then

Years	Century n	Cc	Anchor Day
$1900-1999$, $2300-2399$, $2700-2799$ etc.	$n \equiv 0 \pmod 4$	**3**	Wednesday
$2000-2100$, $2400-2499$, $2800-2899$ etc.	$n \equiv 1 \pmod 4$	**2**	Tuesday
$2100-2199$, $2500-2599$, $2900-2999$ etc.	$n \equiv 2 \pmod 4$	**0**	Sunday
$2200-2299$, $2600-2699$, $3000-3099$ etc.	$n \equiv 3 \pmod 4$	**5**	Friday

This is so because the day before March 1 of the first year of these respective centuries was Wednesday, Tuesday, Sunday, and Friday respectively. These are referred to as *anchor days* for the respective centuries and they repeat every four hundred years. For example the anchor day for the twentieth century years $(1900-1999)$ namely Wednesday (or the number 3) is also the anchor day for the twenty-fourth century $(2300 - 2399)$, or twenty-eighth century $(2700 - 2799)$ etc. Since 1993 belongs to twentieth century $n = 20$ and $Cc = 3$. The following diagram illustrates the description of the Anchor Day/Number for various centuries.

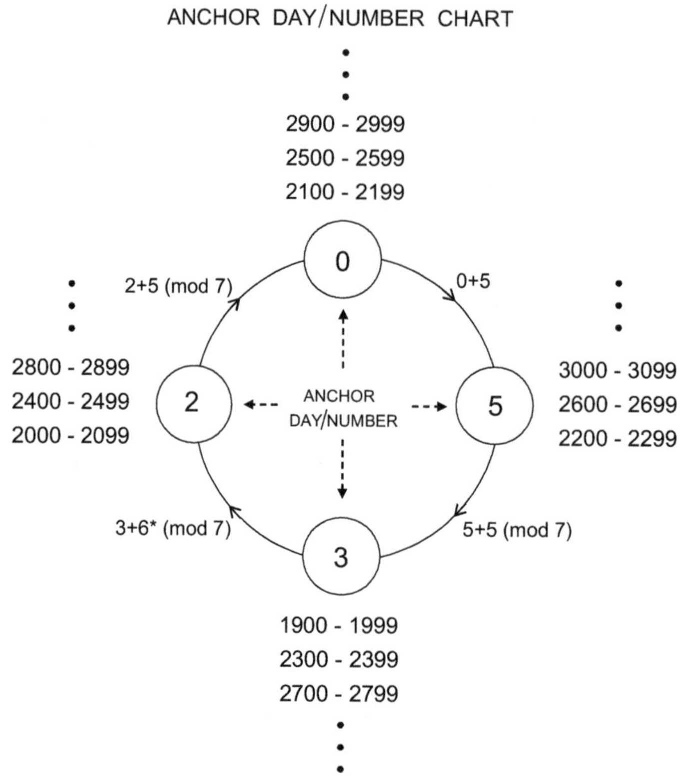

*The years which are multiples of 100 are leap years only when they are multiples of 400.

- **Calculating** Yc

  We note that the quotient of dividing Y by 12 is $\left[\frac{Y}{12}\right]$. Let $Y = 12\left[\frac{Y}{12}\right] + y$, where $0 \le y \le 11$. Then

$$Yc = \left[\frac{Y}{12}\right] + y + \left[\frac{y}{4}\right]$$

  So for the year 1993 we have $Yc = \left[\frac{93}{12}\right] + 9 + \left[\frac{9}{4}\right] = 7 + 9 + 2 = \mathbf{18}$

To calculate the day of the week for December 18, 1993 we add the values of Cc, Yc and DMc for this date. This gives us $3 + \left[\frac{93}{12}\right] + 9 + \left[\frac{9}{4}\right] + 18 - 12 = 3 + 7 + 9 + 2 + 18 - 12 = 27$. As $27 \equiv 6 \pmod 7$, December 18, 1993 was a Saturday.

The following table gives several examples of finding the day of the week using Conway's Doomsday Algorithm as we have described.

Some Examples

Date (mm-dd-yyyy)	d	C	Y	Cc	Yc	DMc	Cc + Yc + DMc	Day of the week
04-05-1956	5	19	56	3	4+8+2	$5-4$	$3+14+1 \equiv 4 \pmod 7$	Thursday
01-26-2976	26	29	76	0	6+4+1	$26-4$	$0+11+22 \equiv 5 \pmod 7$	Friday
01-25-1897	25	18	97	5	8+1	$25-3$	$5+9+22 \equiv 1 \pmod 7$	Monday
07-04-1886	4	19	86	5	7+2	$4-11$	$5+9-7 \equiv 0 \pmod 7$	Sunday
12-18-1993	18	19	93	3	7+9+2	$18-12$	$3+18+6 \equiv 6 \pmod 7$	Saturday
10-05-1869	05	18	69	5	5+9+2	$5-10$	$5+16-5 \equiv 2 \pmod 7$	Tuesday
08-14-1946	14	19	46	3	3+10+2	$14-8$	$3+15+6 \equiv 3 \pmod 7$	Wednesday
01-14-2000	14	20	0	2	0	$14-4$	$2+0+10 \equiv 5 \pmod 7$	Friday

John Conway's Doomsday Algorithm

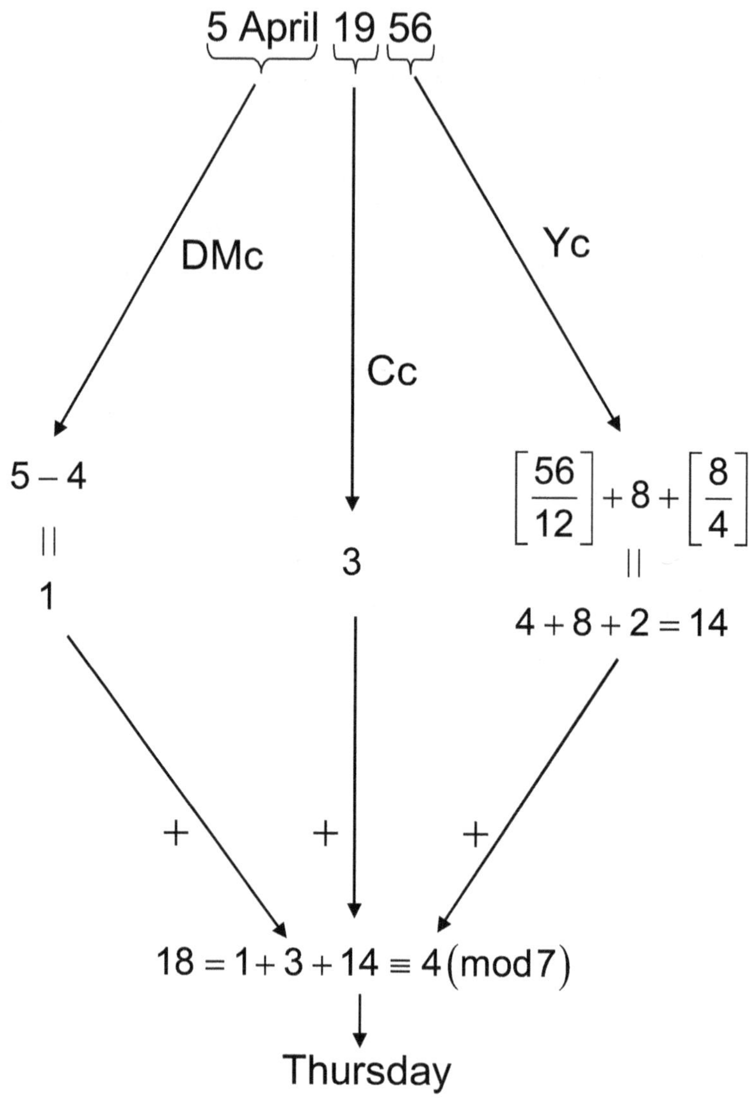

John Conway's Doomsday Algorithm

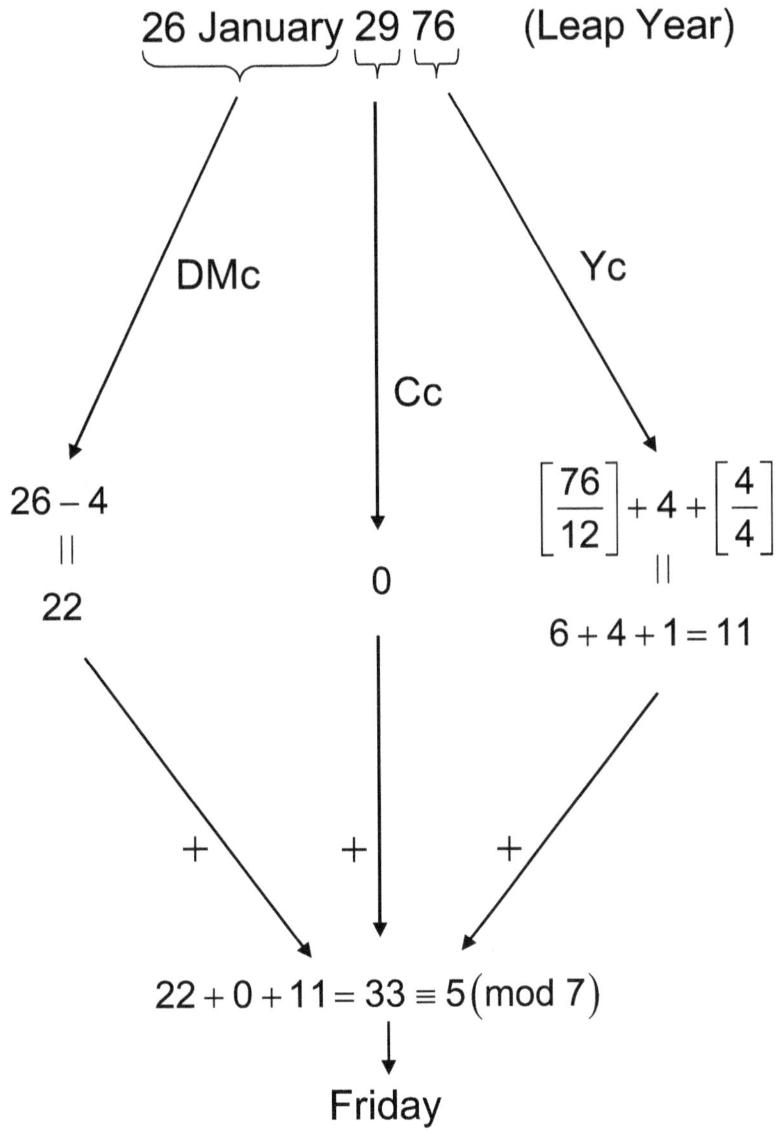

John Conway's Doomsday Algorithm

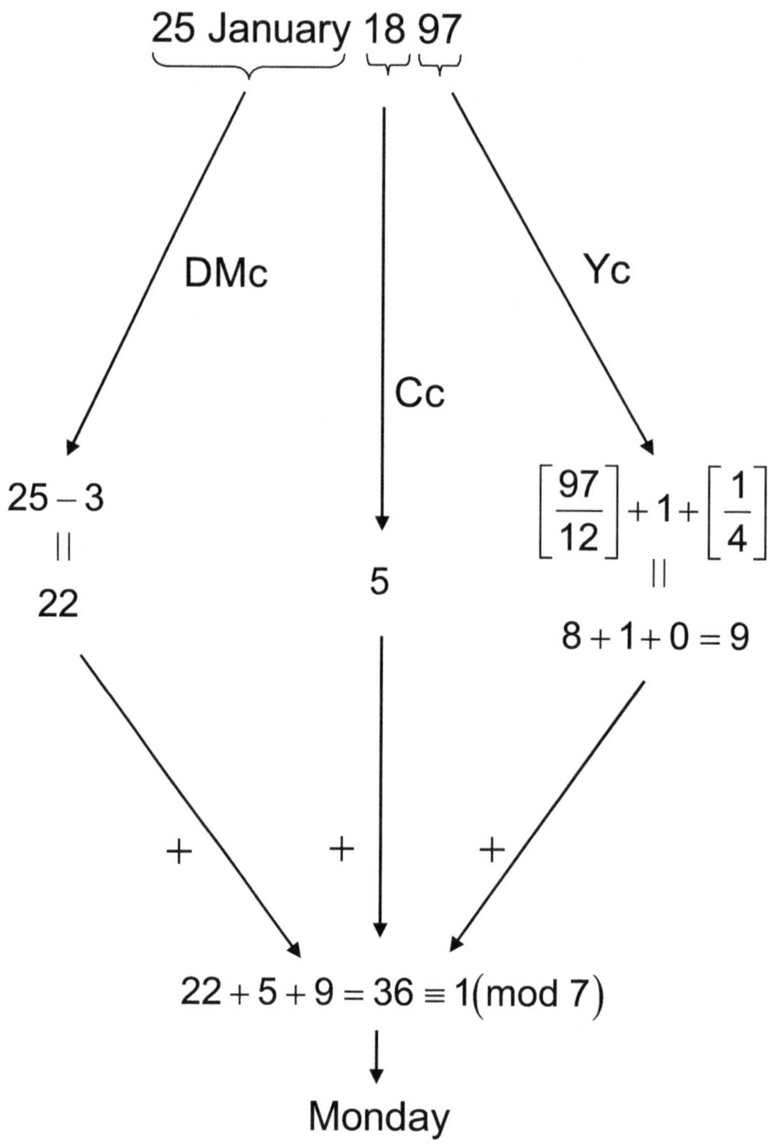

John Conway's Doomsday Algorithm

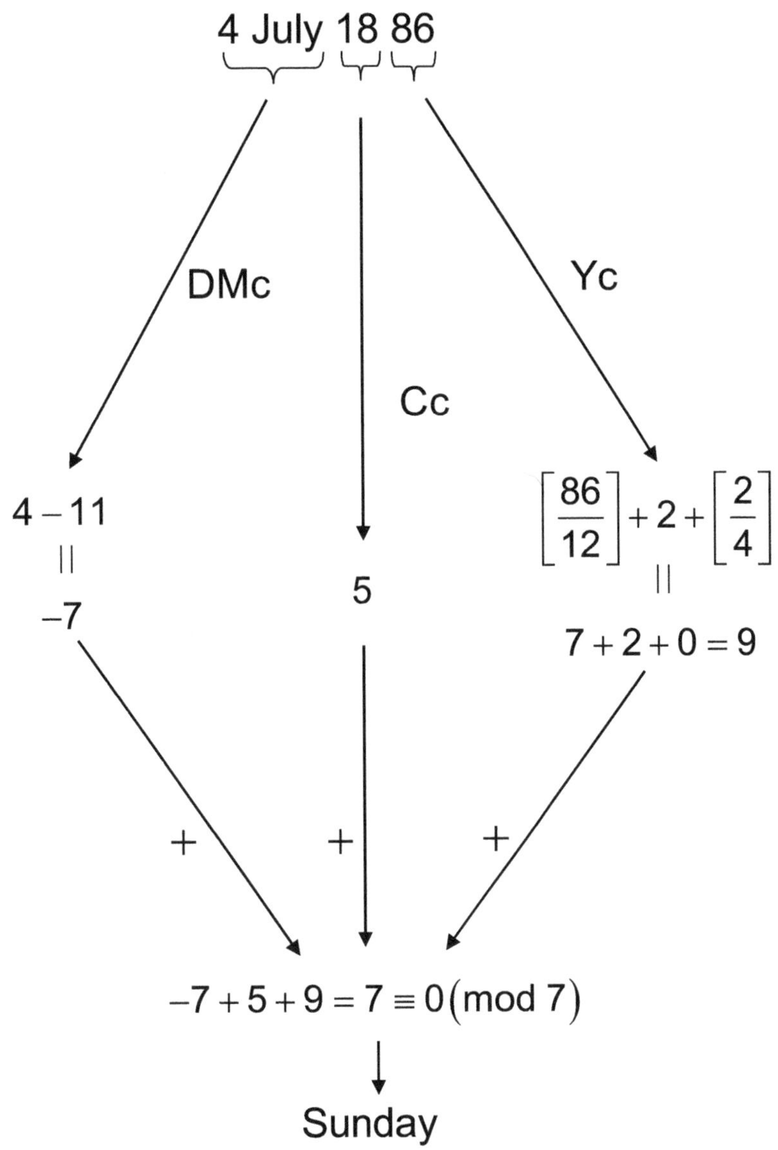

Appendix 9

Cubic Polynomials with Rational Roots and Critical Points

In Calculus I, we teach how to graph a function given by a formula. A natural function for this purpose is a cubic polynomial. To make the problem computationally simple all the "important" points should be rational. In what follows we find all cubic polynomials f with rational coefficients having rational roots such that the roots of f', and f'' are also rational. This question has been addressed to a certain extent in Bruggeman (1980) and Buddenhagen (1992). Our discussion of the same follows Gupta and Szymanski (2010).

Setting Up

First, note that if $f'(x) = (x - r)(x - s)$ with rational r, s then $f''(x) = 2x - r - s$ has the root $\dfrac{r+s}{2}$ which is also rational. Thus we may restrict our attention only to f and f'.
Let us take

$$f(x) = (x^2 + bx + c)(x - d),$$

with real b, c, d, and suppose the quadratic factor has two real roots. This means $\Delta = b^2 - 4c \geq 0$.

Now

$$x^2 + bx + c = \left(x + \frac{b}{2}\right)^2 - \frac{\Delta}{4}$$

and

$$f(x) = \left(\left(x + \frac{b}{2}\right)^2 - \frac{\Delta}{4}\right)\left(\left(x + \frac{b}{2}\right) - \left(\frac{b}{2} + d\right)\right).$$

Letting $y = x + \dfrac{b}{2}$, $u = \dfrac{b}{2} + d$, $v^2 = \dfrac{\Delta}{4}$, we conclude that it is enough to consider

$$f(y) = (y^2 - v^2)(y - u)$$

with rational u, v. Indeed, to get all the roots of f rational, it is enough to assume that u, v, and one of the numbers b or d are rational, see Bruggeman (1980).

The quadratic polynomial

$$f'(y) = 3y^2 - 2yu - v^2$$

has rational roots, if and only if its discriminant $4(u^2 + 3v^2)$ is the square of a rational number. Observe that any pair u, v of *rational* numbers such that $4(u^2 + 3v^2)$ is the square of a *rational* number is a rational multiple of a pair of *integers* u, v such that $u^2 + 3v^2$ is the square of an *integer.*

Thus the key to our problem is to find the integer solutions of the equation

$$u^2 + 3v^2 = w^2. \tag{1}$$

Without loss of generality, we consider only positive u, v, and w.

Primitive Solutions

In this section we find the primitive solutions (i.e., $gcd(u, v, w) = 1$) of (1). First, notice that if u, v, and w are all even then the solution is not primitive. Second, it is not possible that u be even and v, w be odd because in that case $u^2 + 3v^2$ will be congruent to 3 mod 4 whereas w^2 will be congruent to 1 mod 4, and so they cannot be equal.

Therefore we need only consider two cases.

Case I $\quad$ **u is odd, v is even, and w is odd.**

As we are considering primitive solutions, we have $gcd(u, v, w) = 1$. This implies that $gcd(u, v) = gcd(u, w) = gcd(v, w) = 1$. Since v is even, we let $v = 2t$ for some integer t. Thus $3v^2 = 12t^2 = w^2 - u^2 = (w + u)(w - u)$. As both u and w are odd, $w + u$ and $w - u$ are even. Hence $3t^2 = \frac{w+u}{2} \cdot \frac{w-u}{2}$. It follows that 3 divides either $\frac{w+u}{2}$ or $\frac{w-u}{2}$, but not both, as these are relatively prime. Without loss of generality we may assume that 3 divides $\frac{w+u}{2}$. Then

$$t^2 = \frac{1}{3} \cdot \frac{w+u}{2} \cdot \frac{w-u}{2}.$$

As $\frac{1}{3} \cdot \frac{w+u}{2}$ and $\frac{w-u}{2}$ are relatively prime, each must be the square of an integer. We write

$$\frac{1}{3} \cdot \frac{w+u}{2} = m^2, \qquad \text{and} \qquad \frac{w-u}{2} = n^2$$

for some integers m and n. This gives the complete parameterization of Case I:

$$u = 3m^2 - n^2, \qquad v = 2mn, \qquad \text{and} \qquad w = 3m^2 + n^2.$$

Clearly, m and n are relatively prime and have different parity. Since $gcd(u, v, w) = 1$, n is not divisible by 3. As an example, if $m = 2, n = 1$, then $(u, v, w) = (11, 4, 13)$ and $11^2 + 3 \cdot 4^2 = 169 = 13^2$.

Case II $\quad$ **u and v are odd and w is even**

As in Case I, we assume $gcd(u, v) = gcd(u, w) = gcd(v, w) = 1$. We write $3v^2 = w^2 - u^2 = (w + u) \cdot (w - u)$. As before without loss of generality we may assume that 3 divides $w + u$. We then write

$$v^2 = \frac{w+u}{3} \cdot (w - u).$$

Since $\frac{w+u}{3}$ and $(w - u)$ are relatively prime, each must be the square of an integer. We write

$$\frac{w+u}{3} = m^2, \qquad \text{and} \qquad w - u = n^2$$

for some integers m and n. This gives the complete parameterization in case II:

$$u = \frac{3m^2 - n^2}{2}, \qquad v = mn, \qquad \text{and} \qquad w = \frac{3m^2 + n^2}{2}$$

where m and n are relatively prime and are both odd. Since u, v are assumed to be positive, $3m^2 > n^2$. Since $gcd(u, v, w) = 1$, n is not divisible by 3. For example, if $m = 3, n = 1$ then $(u, v, w) = (13, 3, 14)$ and $13^2 + 3 \cdot 3^2 = 14^2$.

Complete Description and Examples

Let us express coefficients of the polynomial f in terms of u, v, and d. It is clear from the introduction that

$$b = 2(u - d), \qquad c = \tfrac{b^2 - 4v^2}{4} = (u - d - v)(u - d + v), \tag{2}$$

hence

$$f(x) = (x^2+bx+c)(x-d) = x^3+(2u-3d)x^2+(u^2-4ud+3d^2-v^2)x-d(u-d-v)(u-d+v). \tag{3}$$

Then

$$f^{'}(x) = 3x^2 + 2(2u - 3d)x + (u^2 - 4ud + 3d^2 - v^2),$$

whose discriminant $4(u^2 + 3v^2)$ is the square of an integer. In particular if we let $u = d$ then

$$f(x) = x^3 - ux^2 - v^2x + uv^2 = (x - u) \cdot (x^2 - v^2). \tag{4}$$

Tables 1 and 2 give some examples of cubic polynomials with the desired properties.

Table 1. The case $u = d$.

u	v	$f(x)$	roots of f	roots of f'
1	1	$x^3 - x^2 - x + 1$	$1, \pm 1$	$1, -\frac{1}{3}$
1	4	$x^3 - x^2 - 16x + 16$	$1, \pm 4$	$-2, \frac{8}{3}$
1	15	$x^3 - x^2 - 225x + 225$	$1, \pm 15$	$9, -\frac{25}{3}$
11	5	$x^3 - 11x^2 - 25x + 275$	$11, \pm 5$	$-1, \frac{25}{3}$
11	4	$x^3 - 11x^2 - 16x + 176$	$16, \pm 4$	$8, -\frac{2}{3}$
13	3	$x^3 - 13x^2 - 9x + 117$	$13, \pm 3$	$9, -\frac{1}{3}$
13	35	$x^3 - 13x^2 - 1225x + 15925$	$13, \pm 35$	$25, -\frac{49}{3}$
61	35	$x^3 - 61x^2 - 1225x + 74725$	$61, \pm 35$	$49, -\frac{25}{3}$

Table 2. The case $u \neq d$.

u	v	d	b	c	$f(x)$	roots of f	roots of f'
1	4	2	-2	-15	$x^3 - 4x^2 - 11x + 30$	$2, -3, 5$	$-1, \frac{11}{3}$
11	4	3	16	48	$x^3 + 13x^2 - 144$	$3, -4, -12$	$0, -\frac{26}{3}$
11	5	2	18	56	$x^3 + 16x^2 + 20x - 112$	$2, -4, -14$	$-10, -\frac{2}{3}$
13	3	5	16	55	$x^3 + 11x^2 - 25x - 275$	$5, -5, -11$	$1, -\frac{25}{3}$
1	15	7	-12	-189	$x^3 - 19x^2 - 105x + 1323$	$7, -9, 21$	$15, -\frac{7}{3}$
13	8	6	14	-15	$x^3 + 8x^2 - 99x + 90$	$6, 1, -15$	$-9, \frac{11}{3}$
23	12	13	20	-14	$x^3 + 7x^2 - 304x + 572$	$13, 2, -22$	$8, -\frac{38}{3}$
11	4	2	18	65	$x^3 + 16x^2 + 29x - 130$	$2, -5, -13$	$-1, -\frac{29}{3}$

Problems

1. Show that if $2^n + 1$ is a prime then n is a power of 2.

   Solution:

   We show that if n is not a power of 2 then $2^n + 1$ is not a prime number. Let $n = k \cdot 2^m$, where $k > 1$ is an odd number. Let $2^{2^m} = x$. Then $2^n + 1 = 2^{k \cdot 2^m} + 1 = x^k + 1$, which is divisible by $x + 1$.

2. Show that if $2^k - 1$ is a prime then k is a prime.

   Solution:

   We show that if k is not a prime then $2^k - 1$ is not a prime. If $k = m \cdot n$ then $2^k - 1$ is divisible by $2^m - 1$ as well as $2^m - 1$.

3. Show that if n is odd then $n^4 + 4n^2 + 11$ is divisible by 16.

   Solution:

   If n is odd then $n^4 \equiv 1 \pmod{16}$ and also $n^2 \equiv 1 \pmod 8$. So $n^4 + 4n^2 + 11 \equiv 1 + 4 + 11 = 16 \equiv 0 \pmod{16}$.

4. Show that if p and $p^2 + 8$ are both prime numbers then $p = 3$.

   Solution:

   If p is a prime and $p \neq 3$ then $p^2 + 8$ is a multiple of 3.

5. Prove that n^5 and n have same last digit.

   Solution:

   It follows from the fact that $n^5 \equiv n \pmod{10}$.

6. Prove that if n is any positive integer then $3^{3n+3} - 26n - 27 \equiv 0 \pmod{169}$

   Solution:

   We have $3^{3n+3} = 27^{n+1} = (1 + 26)^{n+1} \equiv 1 + (n+1)26 = 26n + 27 \pmod{26^2}$. Therefore $3^{3n+3} - 26n - 27 \equiv 0 \pmod{26^2}$ in particular $3^{3n+3} - 26n - 27 \equiv 0 \pmod{169}$.

7. Without actually computing the determinant as it is, show that

$$det \begin{bmatrix} 4771 & 1452 & 8404 & 3275 & 9163 \\ 6573 & 8056 & 7312 & 2266 & 3638 \\ 9712 & 2574 & 4612 & 4321 & 7196 \\ 8154 & 2701 & 6007 & 2147 & 7465 \\ 2158 & 7602 & 5995 & 2327 & 8882 \end{bmatrix} \neq 0$$

Solution:

$$x = det \begin{bmatrix} 4771 & 1452 & 8404 & 3275 & 9163 \\ 6573 & 8056 & 7312 & 2265 & 3639 \\ 9712 & 2574 & 4612 & 4321 & 7196 \\ 8154 & 2701 & 6007 & 2147 & 7465 \\ 2158 & 7602 & 5995 & 2327 & 8882 \end{bmatrix}$$

Working modulo 4 we have

$$x \equiv det \begin{bmatrix} 3 & 0 & 0 & 3 & 3 \\ 1 & 0 & 0 & 1 & 3 \\ 0 & 2 & 0 & 1 & 0 \\ 2 & 1 & 3 & 3 & 1 \\ 2 & 2 & 3 & 3 & 2 \end{bmatrix}$$

Subtracting the first column from the last two columns we have

$$x \equiv det \begin{bmatrix} 3 & 0 & 0 & 0 & 0 \\ 1 & 0 & 0 & 0 & 2 \\ 0 & 2 & 0 & 1 & 0 \\ 2 & 1 & 3 & 1 & 3 \\ 2 & 2 & 3 & 1 & 0 \end{bmatrix}$$

Expanding by first row (twice in succession) we have

$$x \equiv 3 \cdot (-2) \cdot det \begin{bmatrix} 2 & 0 & 1 \\ 1 & 3 & 1 \\ 2 & 3 & 1 \end{bmatrix} \equiv 2 \cdot det \begin{bmatrix} 2 & 0 & 1 \\ 3 & 3 & 0 \\ 0 & 3 & 0 \end{bmatrix} \equiv 2 \pmod 4$$

Therefore x cannot be zero.

8. Show that if n is a natural number then

$$1+\frac{1}{3}+\frac{1}{5}+\frac{1}{7}\cdots+\frac{1}{2n+1}$$

is not an integer.

Solution:

Let k be the largest natural number such that $3^k \leq 2n+1$. Let

$$N = LCM\{1, 3, 5, \cdots, 2n+1\}.$$

Then $N = 3^k \cdot m$, where $m \geq 1$ is an odd natural number which is *not* divisible by 3.

Let

$$x = 1+\frac{1}{3}+\frac{1}{5}+\frac{1}{7}\cdots+\frac{1}{2n+1}.$$

If x is an integer then so is $x \cdot N$ and the later is evidently a multiple of 3. We observe that the denominator of only one term in the expression for x (namely $\frac{1}{3^k}$) is divisible by 3^k. Therefore in

$$x \cdot N = \frac{N}{1}+\frac{N}{3}+\frac{N}{5}+\frac{N}{7}\cdots+\frac{N}{2n+1}$$

each term in the RHS except one (namely $\frac{N}{3^k}$) is divisible by 3. So the RHS is not a multiple of 3. So our assumption that x is an integer leads to a contradiction.

9. Find the remainder when the number X in the first column is divided by the number Y in the second column.

X	Y	$X \pmod{Y}$
23^{243}	17	
3^{100}	91	

Answer:

X	Y	$X \pmod{Y}$
23^{243}	17	12
3^{100}	91	81

10. Let $N = 7^{11}$, which is a 10-digit number. When we represent N in base nine that too turns out to be a 10-digit number. In the base nine representation of N, all the

nine digits $\{0, 1, 2, 3, 4, 5, 6, 7, 8, \}$ appear and one of the digits appear twice. Find the digit which appears twice.

Solution:

$N = 7^{11} \equiv 7 \pmod 8$. Let x represent the repeated digit in the base nine representation of N. We have

$$\begin{aligned} N &= (b_9b_8b_7b_6b_5b_4b_3b_2b_1b_0)_{\text{nine}} \text{ where } b_i \in \{0, 1, 2, 3, 4, 5, 6, 7, 8, \} \\ &= = \textstyle\sum_{i=0}^{i=9} b_i 9^i \\ &= \textstyle\sum_{i=0}^{i=9} b_i (8+1)^i \\ &\equiv \textstyle\sum_{i=0}^{i=9} b_i \pmod 8 \\ &\equiv x + \textstyle\sum_{i=0}^{i=8} i \pmod 8 \\ &\equiv x + 36 \pmod 8 \end{aligned}$$

So $x + 36 \equiv 7 \pmod 8$, which implies that $x = 3$.

11. Let $N = 7^p - 6^p - 1$ where $p > 3$ is a prime. Prove that $N \equiv 0 \pmod{43}$.

Solution:

First note that order of 7 mod 43 is 6 and order of 6 mod 43 is 3. Also $7 \times 37 \equiv 1 \pmod{43}$ or $7^{-1} \equiv 37 \pmod{43}$. As the prime p is greater than 3, we have $p \equiv 1, 5 \pmod 6$.

If $p = 6k + 1$ then we have

$$N = 7^{6k+1} - 6^{6k+1} - 1 \equiv 7 - 6 - 1 = 0 \pmod{43}.$$

If $p = 6k + 5$ then we have

$$N = 7^{6k+5} - 6^{6k+5} - 1 \equiv 7^{-1} - 6^2 - 1 \equiv 37 - 36 - 1 = 0 \pmod{43}.$$

12. Let $n \in \mathbb{N}$. Prove that if n is not divisible by 3 then $2^n(2^n + 1) + 1$ is not a prime number.

Solution:

Let $N = 2^n(2^n + 1) + 1 = 4^n + 2^n + 1$. If $n = 3k + 1$ then we have

$$4^n + 2^n + 1 = 4^{3k+1} + 2^{3k+1} + 1 \equiv 4 + 2 + 1 \equiv 0 \pmod 7.$$

If $n = 3k + 2$ then we have

$$4^n + 2^n + 1 = 4^{3k+2} + 2^{3k+2} + 1 \equiv 16 + 4 + 1 \equiv 0 \pmod 7.$$

13. Prove that $11^{4n+2} + 19^{4n+2} \equiv 0 \pmod{241}$.

Solution:

We observe that $11^2 = 121 \equiv -120 \pmod{241}$ and $19^2 = 361 \equiv 120 \pmod{241}$. Therefore $11^4 \equiv 19^4 \equiv 120^2 \pmod{241}$. Thus we have

$$11^{4n+2} + 19^{4n+2} \equiv 120^n \cdot 11^2 + 120^n \cdot 19^2 = 120^n(121 + 361) = 120^n \cdot 482 \equiv 0 \pmod{241}.$$

14. Prove that for $n \geq 0$, $3^{2n+1} + 2^{n+2} \equiv 0 \pmod{7}$.

Solution:

$$3^{2n+1} + 2^{n+2} = (3^2)^n \cdot 3 + 2^{n+2} \equiv 2^n \cdot 3 + 2^{n+2} = 2^n(3+4) \equiv 0 \pmod{7}.$$

15. Prove that for $n \geq 0$, $3^{n+2} + 4^{2n+1} \equiv 0 \pmod{13}$.

Solution:

$$3^{n+2} + 4^{2n+1} \equiv 3^{n+2} + 3^n \cdot 4 = 3^n(3^2 + 4) \equiv 0 \pmod{13}.$$

16. Find (all) values for the digit X such that $48^{499\mathbf{X}579} \equiv 9 \pmod{23}$.

Solution:

We note that $48 \equiv 2 \pmod{23}$ and $9 \equiv 32 = 2^5 \pmod{23}$, and $2^{11} \equiv 1 \pmod{23}$. As $499X579 \equiv 27 - (16 + X) = 11 - X \equiv -X \pmod{11}$ we have $2^{499\mathbf{X}579} \equiv 2^{-X} \pmod{23}$. Therefore

$$2^{-X} \equiv 2^5 \pmod{23} \Longrightarrow 2^{X+5} \equiv 1 \pmod{23} \Longrightarrow X \equiv 6 \pmod{11}.$$

So $X = 6$.

17. Find the last digit of $7^{139} + 13^{2001}$.

Solution:

$$7^{139} + 13^{2001} \equiv 7^3 + 3 \equiv 6 \pmod{10}.$$

18. Prove that $20^{15} - 1 \equiv 0 \pmod{20801}$.

Solution:

Since $20801 = 11 \cdot 31 \cdot 61$, it would suffice to show the followings:

(a) $20^{15} - 1 \equiv 0 \pmod{11}$.

(b) $20^{15} - 1 \equiv 0 \pmod{31}$.

(c) $20^{15} - 1 \equiv 0 \pmod{61}$.

(a) Working modulo 11 (and using Fermat's Little Theorem) we have

$20^{15} \equiv 20^5 \equiv (-2)^5 = -32 \equiv 1 \pmod{11}$.

(b) Working modulo 31 we have

$20^2 \equiv (-11)^2 = 121 \equiv -3 \pmod{31}$. So $20^4 \equiv 9 \pmod{31}$ and thus $20^5 \equiv -6 \pmod{31}$, and $20^{10} \equiv 5 \pmod{31}$. This gives $20^{15} \equiv 1 \pmod{31}$.

(c) Working modulo 61 we have

$20^2 = 400 \equiv 34 \pmod{61}$, therefore $20^3 \equiv 680 \equiv 9 \pmod{61}$. Squaring we get $20^6 \equiv 20 \pmod{61}$. Thus $20^5 \equiv 1 \pmod{61}$. Hence $20^{15} \equiv 1 \pmod{61}$.

19. A class in number theory was to divide itself into groups of equal sizes for the purpose of groupstudy. When the class was divided into groups of three, two students were left out; when divided into groups of four, one was left out. When it was divided into groups of five the students found that if the professor was added to one of the groups, no one was left out. How many students were there in the class. (*Note:* The college rules do not allow a class size of more than 50.)

Solution:

Let X denote the number of students in the class. So we have $X \leq 50$ and, $X \equiv 2 \pmod{3}$, $X \equiv 1 \pmod{4}$ and $X + 1 \equiv 0 \pmod{5}$. A simple calculation shows that $X \equiv 29 \pmod{60}$. Since $X \leq 50$, we must have $X = 29$.

20. Without doing the actual calculations find the remainder when

$$1^5 + 2^5 + 3^5 + \cdots + 99^5 + 100^5$$

is divided by 16.

Solution:

We first note that if n is odd then $n^4 \equiv 1 \pmod{16}$ and so $n^5 \equiv n \pmod{16}$. Therefore

$$1^5 + 2^5 + 3^5 + \cdots + 99^5 + 100^5 \equiv 1 + 3 + 5 + 7 + \cdots + 97 + 99 \pmod{16}.$$

The sum of first four odd natural numbers is a multiple of 16 and so is the sum of next four odd numbers and so on. Therefore, the given sum is congruent to $97 + 99 \equiv 4 \pmod{16}$.

21. Show that if x and y are integers then $\dfrac{x^2 - 2}{2y^2 + 3}$ is never an integer.

Solution:

It is enough to show that there is a prime number p which divides $2y^2+3$ but does not divide x^2-2. As $2y^2+3 \equiv \pm 3 \pmod 8$, $2y^2+3$ has a prime divisor $p \equiv \pm 3 \pmod 8$. However any odd prime divisor of x^2-2 is of the form $8k \pm 1$.

22. A five digit positive integer X is such that the last five digits of its square X^2 are same as those of X (in the same order). Find all such possible X.

Solution: We have $X(X-1) \equiv 0 \pmod{10^5}$. Since X and $X-1$ are relatively prime, two cases arise:

Case I $X \equiv 0 \pmod{2^5}$ and $X-1 \equiv 0 \pmod{5^5}$. This implies that $X \equiv 9376 \pmod{10^5}$. But such an X is not a five-digit number.

Case II $X \equiv 0 \pmod{5^5}$ and $X-1 \equiv 0 \pmod{2^5}$. This implies that $X \equiv 90625 \pmod{10^5}$. This gives $X = 90625$.

23. Find the *largest* possible divisor N which will leave the same remainder when divided into the three numbers: 1059, 1417, 2312.

Solution:

We have $1417 - 1059 = 358 \equiv 0 \pmod N$. $2312 - 1059 = 1253 \equiv 0 \pmod N$. $2312 - 1417 = 8950 \pmod N$. Therefore $N = GCD(358, 1253, 895) = 179$.

24. Find the digits X and Y such that the number $XY2130117169303960693$ is divisible by 99.

Solution:

By direct calculation, we have $X + Y \equiv 2 \pmod 9$ and $X - Y \equiv 1 \pmod{11}$. Since X and Y are digits, the only solution is $X = 6$ and $Y = 5$.

25. Find the least positive integer N satisfying the conditions that $N/2$ is a square, $N/3$ is a cube, and $N/5$ is a fifth power.

Solution:

Let $N = 2^a 3^b 5^c$. The given conditions imply that

$$\begin{aligned}
a &\equiv 1 \pmod 2, & a &\equiv 0 \pmod 3, & a &\equiv 0 \pmod 5 \Longrightarrow a = 15.\\
b &\equiv 0 \pmod 2, & b &\equiv 1 \pmod 3, & b &\equiv 0 \pmod 5 \Longrightarrow b = 10.\\
c &\equiv 0 \pmod 2, & c &\equiv 0 \pmod 3, & c &\equiv 1 \pmod 5 \Longrightarrow c = 6.
\end{aligned}$$

These give $N = 2^{15}3^{10}5^6 = 30,233,088,000,000$.

26. Amy knew that her mathematician grandmother is no more than 80 years old but wanted to know her correct age. She asked her grandmother the year or her birth. The grandmother replied that in the year x^2, she was x years old. Can you help Amy find out her grandmother's year of birth and her age at the time of conversation which took place in the year 1970?

Solution:

Since in 1970 grandmother is no more than 80 years old, we have $1890 < x^2 < 1970$. This implies that $x = 44$. So in the year 1936, the grandmother was 44 years old so she was born in 1892 and so she was 78 years old in 1970.

27. A man received a cheque for a certain amount of money, but upon cashing it the teller mistook the number of dollars for the number of cents and conversely. Not noticing this, the man spent 68 cents and discovered to his surprise that he had twice as much money as the cheque was originally written for. Determine the smallest amount of money for which the cheque could have been written.

Solution:

Suppose that the cheque was written for x dollars and y cents. Since the teller mistook dollars for cents and vice versa both x and y are less than 100. The given condition implies that $2(100x + y) = 100y + x - 68$, or $199x - 98y + 68 = 0$. This gives $x \equiv 10 \pmod{98}$. Hence we have $x = 10$ and $y = 21$.

28. Note that 1999 is a prime number. Fill in the following table giving orders (modulo 1999) of the numbers in the first column:

x	Order of x modulo 1999
2	
3	
5	
7	

Solution:

x	Order of x modulo 1999
2	333
3	1998
5	999
7	666

29. Let $N = 722$. Fill in the following table giving the orders (modulo 722) of the numbers in the first column.

x	Order of x modulo 722
3	
5	
11	
13	

Solution:

x	Order of x modulo 722
3	342
5	171
11	57
13	342

30. A professor assigned to teach a course in number theory told the chairperson of the department that he will teach the course only if the number of students enrolled (say X) is such that if it is divided by 3 the remainder is 2, if it is divided by 5 the remainder is 3, and if it is divided by 7 the remainder is 5. Find the minimum

enrollment (i.e., the minimum value of X) for the class before the professor could consent to teach the course.

Answer: 68

31. Fill in the second column which gives the last three digits of the number in the first column.

X	Last three digits of X
17^{9999}	
3^{1000}	
7^{1620}	

32. It is known that a natural number N can be expressed as a sum of three squares if and only if it is *not* of the form $4^a(8k+7)$. Assuming this show that every natural number N can be expressed as $x^2+y^2+z^2$ or $2x^2+y^2+z^2$. In other words, although every natural number can be expressed as a sum of four squares, for example, $N = a^2+b^2+c^2+d^2$ there is no natural number which *necessarily* requires four different nonzero numbers $a, b, c, d \in \mathbb{N}$ in its expression as a sum of four squares.

Solution:

First, we show an odd natural number n can be expressed as $x^2+y^2+2z^2$.

If n is odd then $2n \equiv 2 \pmod 4$ and so $2n$ can be expressed as a sum of three squares. Let $2n = x_1^2+y_1^2+z_1^2$. Since $2n \equiv 2 \pmod 4$, two of the numbers x_1, y_1, z_1 are odd and one even. Without loss of generality, we may assume that x_1, y_1 are odd and z_1 even. Let

$$x = \frac{x_1+y_1}{2}, \qquad y = \frac{x_1-y_1}{2}, \qquad z = \frac{z_1}{2}.$$

Then

$$x^2+y^2+2z^2 = \left(\frac{x_1+y_1}{2}\right)^2 + \left(\frac{x_1-y_1}{2}\right)^2 + 2\left(\frac{z_1}{2}\right)^2 = \frac{x_1^2+y_1^2+z_1^2}{2} = n.$$

Now let N be any natural number. If $N \neq 4^a(8k+7)$ then N is a sum of three squares. If $N = 4^a(8k+7)$ then

$$N = 4^a(8k+7) = 4^a \cdot n = 4^a(x^2+y^2+2z^2) = (2^a x)^2 + (2^a y)^2 + 2(2^a z)^2.$$

33. Find the *smallest* positive integer N satisfying each of the following conditions:

- When N is divided by 2 the remainder is 1.
- When N is divided by 3 the remainder is 2.
- When N is divided by 4 the remainder is 3.
- When N is divided by 5 the remainder is 4.
- When N is divided by 6 the remainder is 5.
- When N is divided by 7 the remainder is 6.
- When N is divided by 8 the remainder is 7.
- When N is divided by 9 the remainder is 8.
- When N is divided by 10 the remainder is 9.
- When N is divided by 11 the remainder is 10.
- When N is divided by 12 the remainder is 11.
- When N is divided by 13 the remainder is 0.
- When N is divided by 14 the remainder is 13.
- When N is divided by 15 the remainder is 14.
- When N is divided by 18 the remainder is 17.
- When N is divided by 20 the remainder is 19.

Solution:

The given conditions imply that it is sufficient to determine the least N satisfying the following congruence:

- $N \equiv 4 \pmod 5$ or $N \equiv -1 \pmod 5$
- $N \equiv 6 \pmod 7$ or $N \equiv -1 \pmod 7$
- $N \equiv 7 \pmod 8$ or $N \equiv -1 \pmod 8$
- $N \equiv 8 \pmod 9$ or $N \equiv -1 \pmod 9$
- $N \equiv 10 \pmod{11}$ or $N \equiv -1 \pmod{11}$
- $N \equiv 0 \pmod{13}$

This means that $N \equiv -1 \pmod{5 \cdot 7 \cdot 8 \cdot 9 \cdot 11}$ that is $N \equiv -1 \pmod{27720}$ and $N \equiv 0 \pmod{13}$. The former congruence implies that $N = -1 + 27720k$ for some integer k. So we are left to solve the congruence

$$-1 + 27720k \equiv 0 \pmod{13}.$$

This gives $k \equiv 10 \pmod{13}$. Let $k = 10 + 13t$ for some t. Substituting this value of k in $N = -1 + 27720k$ we get $N = 277199 + 360360t$ where t is any integer. So the least positive solution (which corresponds to $t = 0$) is $N = 277199$.

34. Suppose that a child has \$4.55 in his piggy bank which consists of quarters and dimes. What is the maximum and minimum number of total coins in the child's piggy bank.

Solution:

Let x denote the number of quarters and y the number of dimes. Then we have the Diophantine equation $25x + 10y = 455$ or $5x + 2y = 91$. Considering the later equation modulo 5, we have $y \equiv 3 \pmod 5$. Let $y = 3 + 5t$ and so $x = 17 - 2t$. Since $x, y \geq 0$, we have $17 - 2t \geq 0$, or $t \leq \frac{17}{2}$ and $3 + 5t \geq 0$, or $t \geq -\frac{3}{5}$. Therefore $-\frac{3}{5} \leq t \leq \frac{17}{2}$ and as t is an integer we have $0 \leq t \leq 8$. The following table gives the values of x and y for various values of t in the range $0 \leq t \leq 8$:

t	0	1	2	3	4	5	6	7	8
x	17	15	13	11	9	7	5	3	1
y	3	8	13	18	23	28	33	38	43

The maximum value of $x + y$ is 44 and the minimum value of $x + y$ is 20.

35. Prove that the Diophantine equation $x^2 - 2y^2 + 8z = 19$ has no solutions.

Proof: Suppose that Diophantine equation $x^2 - 2y^2 + 8z = 19$ has solutions. Evidently, x cannot be even. If x is odd then $x^2 \equiv 1 \pmod 8$ which would imply that $-2y^2 \equiv 2 \pmod 8$, which is not possible.

36. Let $N = 92815xyz$ where $x, y, z \in \{0, 1, 2, 3, 4, 5, 6, 7, 8, 9, \}$ be a 8-digit decimal number which is divisible by 693. Find the digits x, y, z.

Solution:

We have

$N \equiv 0 \pmod{11} \Longrightarrow x - y + z \equiv 8 \pmod{11}$.

$N \equiv 0 \pmod 9 \Longrightarrow x + y + z \equiv 2 \pmod 9$.

$N \equiv 0 \pmod 7 \Longrightarrow 2x + 3y + z \equiv 2 \pmod{11}$.

These give $x = 5$, $y = 6$, and $z = 9$ as the only solution.

37. Find the last three digits of the number 11^{9598}.

Solution:

Let $x = 11^{9598}$. Evidently $x \equiv 1 \pmod 8$, and $x = 11^{9598} \equiv x^{98} \pmod{125}$ which implies that $121x \equiv 1 \pmod{125}$ and therefore $x \equiv 31 \pmod{125}$. Using the Chinese Remainder Theorem, we have $x = 281 \pmod{1000}$.

38. When a customer cashed a cheque for x dollars and y cents at a bank the teller by mistake gave him y dollars and x cents. The customer found that he got 10 cents more than twice the proper amount. For how much was the cheque written?

Solution:

Suppose that the cheque was written for x dollars and y cents. Then $0 \leq y \leq 99$. The given conditions imply that $100y + x = 2(100x + y) + 10$, that is, $199x - 98y = -10$. Considering the last equation modulo 199 we have $-98y \equiv -10 \pmod{199}$. This gives $y \equiv 63 \pmod{199}$. Therefore $y = 63$ and $x = 31$. So the cheque was written for 31 dollars and 63 cents.

39. Knowing that the GCD of the three numbers $90ABC17$, $79ABC$, and $491ABC4$ is more than one, determine the digits A, B, and C.

Solution:

Let $GCD(90ABC17, 79ABC, 491ABC4) = N > 1$. Clearly N divides the two numbers

$$90ABC17 - 79ABC \times 100 = 1100017 \text{ and } 491ABC4 - 79ABC \times 10 = 4210004.$$

Since $GCD(1100017, 4210004) = 547$ is a prime number we must have $N = 547$. Let $79ABC = 547 \cdot q$. Now we have

$$79000 < 79ABC < 79999 \text{ or } 79000 < 547 \cdot q < 79999$$

which gives $144.4 < q < 146.25$. Since q is a whole number we have $q = 145$ or $q = 146$. So either $79ABC = 79315$ or $79ABC = 79862$. So either $(A, B, C) = (3, 1, 5)$ or $(A, B, C) = (8, 6, 2)$.

40. Find the last three digits of 7^{9999}.

Solution:

First, we note that $1000 \equiv -1 (\text{mod } 7)$. Since $1000 = -1 + 7 \cdot 143$ we have $7 \cdot 143 \equiv 1 \pmod{1000}$. Or $7^{-1} \equiv 143 (\text{mod } 1000)$. Since $\phi(1000) = 400$, using Fermat's Little Theorem we have $7^{400} \equiv 1 \pmod{1000}$. Or $7^{10000} = (7^{40})^{250} \equiv 1 \pmod{1000}$.

Now computing modulo 1000 we have

$$7^{9999} = 7^{10000} \cdot 7^{-1} \equiv 7^{-1} \equiv 143 \pmod{1000}.$$

Thus, the last three digits of 7^{9999} are 143.

41. Find the last three digits of 2^{5620}.

Let $x = 2^{5620}$. Since $1000 = 2^3 \cdot 5^3$, we find $x(\text{mod } 8)$ and $x(\text{mod } 125)$ and then determine $x(\text{mod } 1000)$ using the Chinese Remainder Theorem. Evidently $x \equiv 0(\text{mod } 8)$. Since $\phi(125) = 100$, we have $2^{5620} = (2^{100})^{56} \cdot 2^{20} \equiv 2^{20}(\text{mod } 125)$. Now,

$2^{10} = 1024 = 1000 + 24$, so $2^{20} \equiv 24^2 \equiv 76 \pmod{125}$.

So $x \equiv 76(\text{mod } 125)$. An easy application of Chinese Remainder Theorem gives $x \equiv 576(\text{mod } 1000)$. Hence, the last three digits of x are 576.

42. Show that no square (of a natural number) has a decimal expansion ending in 79. More generally, find all possible two-digit endings for squares of natural numbers.

Solution:

Since $79 \equiv 3(\text{mod } 4)$ the square of a number cannot end in 79 as square of a number is congruent to 0 or 1 (mod 4). Also since the square of a number is congruent to $0, 1$, or 4 (mod 5), it follows that square of a number is congruent to one of the following numbers (mod 25): 0, 1, 6, 11, 16, 21, 4, 9, 14, 19, 24.

Let x be any natural number. The following table is self-explanatory:

$x^2(\text{mod } 4)$	$x^2(\text{mod } 25)$	$x^2(\text{mod } 100)$	$x^2(\text{mod } 4)$	$x^2(\text{mod } 25)$	$x^2(\text{mod } 100)$
0	0	00	1	0	25
0	1	76	1	1	01
0	4	04	1	4	29
0	6	56	1	6	81
0	9	84	1	9	09
0	11	36	1	11	61
0	14	64	1	14	89
0	16	16	1	16	41
0	19	44	1	19	69
0	21	96	1	21	21
0	24	24	1	24	49

The (two digit) numbers in the third and sixth columns are the possible two-digits endings for integral squares.

43. Using the Chinese Remainder Theorem or otherwise find the least positive X satisfying (simultaneously) the following three congruence:

$$\begin{aligned} 2X &\equiv 3 \pmod 7 \\ 3X &\equiv 2 \pmod{11} \\ 11X &\equiv 6 \pmod{13} \end{aligned}$$

Answer: $X = 712.$

44. Let $a, b \in \mathbb{Z}$, with $gcd(a, b) = 1$. Let p be an odd prime. Prove that if a prime q divides $a + b$ and $\dfrac{a^p + b^p}{a + b}$ then $q = p$.

Proof:

As $a + b \equiv 0 \pmod q$, $b \equiv -a \mod q$. Since $\dfrac{a^p + b^p}{a + b} \equiv 0 \pmod q$, we have

$$\begin{aligned} \frac{a^p + b^p}{a + b} &= \textstyle\sum_{i=1}^{i=p} a^{p-i} b^{i-1} (-1)^{i-1} \\ &\equiv \textstyle\sum_{i=1}^{i=p} a^{p-i} (-a)^{i-1} (-1)^{i-1} \pmod q \\ &\equiv \textstyle\sum_{i=1}^{i=p} a^{p-i} (a)^{i-1} \pmod q \\ &\equiv \textstyle\sum_{i=1}^{i=p} a^{p-1} \pmod q \\ &\equiv p \cdot a^{p-1} \pmod q. \end{aligned}$$

Therefore $q = p$.

45. Let $p > 3$ be a prime number and

$$\sum_{i=1}^{p-1} \frac{1}{i} = 1 + \frac{1}{2} + \frac{1}{3} + \cdots\cdots + \frac{1}{p-1} = \frac{a}{b}$$

where $gcd(a, b) = 1$. Prove that a is divisible by p^2.

Solution:

Evidently

$$\sum_{i=1}^{p-1} \frac{1}{i} = \frac{\sum_{i=1}^{p-1} \frac{(p-1)!}{i}}{(p-1)!}.$$

Since p does not divide $(p-1)!$, it is enough to show that

$$\sum_{i=1}^{i=p-1} \frac{(p-1)!}{i} \equiv 0 \pmod{p^2}.$$

Since $x^{p-1} \equiv (x-1)(x-2)\cdot\cdots\cdot(x-p+1) \pmod{p}$ we have

$$(x-1)(x-2)\cdot\cdots\cdot(x-\overline{p-1}) = x^{p-1} + a_1x^{p-2} + a_2x^{p-3} + \cdots + a_{p-2}x + a_{p-1}$$

where $a_i \equiv 0 \pmod{p}$ for $1 \le i \le p-2$, $a_{p-1} = (p-1)!$, and

$$a_{p-2} = -\sum_{i=1}^{i=p-1} \frac{(p-1)!}{i}.$$

Substituting $x = p$ in

$$(x-1)(x-2)\cdot\cdots\cdot(x-\overline{p-1}) = x^{p-1} + a_1x^{p-2} + a_2x^{p-3} + \cdots + a_{p-2}x + a_{p-1}$$

we have

$$(p-1)! = p^{p-1} + a_1p^{p-2} + a_2p^{p-3} + \cdots + a_{p-3}p^2 + a_{p-2}p + a_{p-1}.$$

Since $a_{p-1} = (p-1)!$, we have

$$p^{p-1} + a_1p^{p-2} + a_2p^{p-3} + \cdots + a_{p-3}p^2 + a_{p-2}p = 0.$$

Since $p > 3$ and $a_i \equiv 0 \pmod{p}$ for $1 \le i \le p-2$ all but the last term in $p^{p-1}+a_1p^{p-2}+a_2p^{p-3}+\cdots+a_{p-3}p^2+a_{p-2}p$ is divisible by p^3 we must have $a_{p-2}\cdot p \equiv 0 \pmod{p^3}$ or $a_{p-2} \equiv 0 \pmod{p^2}$. But as

$$a_{p-2} = -\sum_{i=1}^{i=p-1} \frac{(p-1)!}{i}$$

therefore

$$\sum_{i=1}^{i=p-1} \frac{(p-1)!}{i} \equiv 0 \pmod{p^2}$$

and hence p^2 divides a.

Note: The statement of this problem is known as a theorem of Wolstenholme, see Hardy and Wright (1960).

46. Show that if $19a^2 \equiv b^2 (\text{mod } 7)$ then $19a^2 \equiv b^2 (\text{mod } 7^2)$.

Solution:

Since $\left(\frac{19}{5}\right) = \left(\frac{5}{7}\right) = -1$, 19 is a quadratic nonresidue mod 7, b and a must be multiples of 7. So a^2 and b^2 are multiples of 7^2.

47. Find the least positive integer k such that $5^k \equiv 1 \pmod{163}$.

Solution:

As $\left(\frac{5}{163}\right) = -1$, $5^{81} \not\equiv 1 \pmod{163}$ and so k must be even. Since $5^6 \equiv 140 \pmod{163}$ and $5^{18} \equiv 58 \pmod{163}$, we have $k = 54$.

48. Prove 6 cannot be expressed as a square of two rational numbers.

Solution:

It follows from the fact that the Diophantine equation has $x^2 + y^2 = 6z^2$ has no solutions.

49. Prove that n divides $2^{n!} - 1$, if n is odd.

Hint:

Show that if n is odd then $\phi(n)$ divides $n!$.

50. Find all natural numbers n such that $\phi(n) = \frac{n}{2}$.

Answer: $n = 2^k$, k any positive integer.

51. Find all natural numbers n such that $\phi(n) = \phi(2n)$.

Answer: n must be odd.

52. Find all natural numbers n such that $\phi(n) = \frac{n}{3}$.

Answer: $n = 2^a 3^b$, $a, b > 0$.

53. Find all natural numbers N such that $\phi(N) = 12$.

Solution:

Evidently, N cannot be divisible by any prime greater than 13. A case by case examination yields that N can only be one of following numbers $13, 21, 26, 28, 36,$ and 43 as illustrated in the following diagram (note that N cannot be divisible by 5 as $\phi(5) = 4$ and $\phi(n) = 3$ has not solutions).

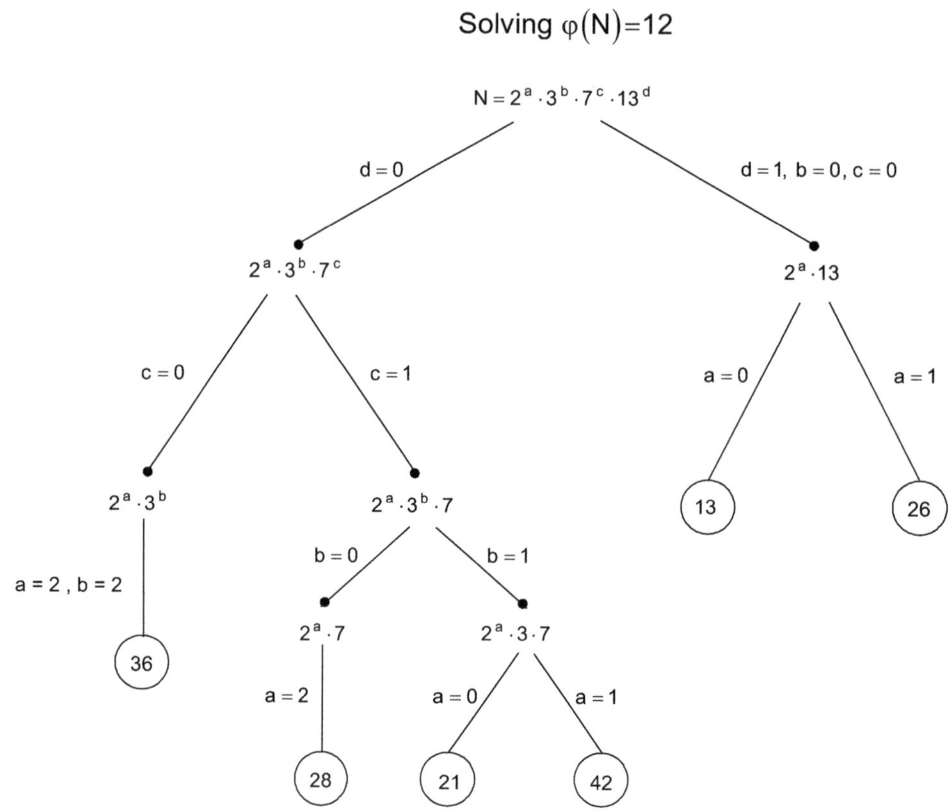

54. Prove that the sum of all positive integers less than n and prime to n is $\dfrac{n\phi(n)}{2}$.

Solution:

This has been discussed in Chapter 2.

55. Find all integers N such that $\phi(N)$ divides N.

Solution: Clearly $N = 1$ satisfies the given condition. So let us assume that $N > 1$. Let

$$N = p_1^{a_1} p_2^{a_2} \cdots p_k^{a_k}$$

where $p_1 < p_2 < \cdots < p_k$ are prime numbers. We have

$$\frac{N}{\phi(N)} = \frac{p_1 \cdot p_2 \cdots p_k}{(p_1 - 1) \cdot (p_2 - 1) \cdots (p_k - 1)}$$

Clearly then N must be even and at most one of the numbers $p_1 - 1, p_2 - 1, \cdots, p_k - 1$ can be even. This means that N is even and can have at most one odd prime divisors. So we may assume that $N = 2^a p^b$, where p is an odd prime, $a > 0$

and $b \geq 0$. We then have

$$\frac{N}{\phi(N)} = \frac{2^a \cdot p^b}{2^{a-1} \cdot p^{b-1}(p-1)} = \frac{2 \cdot p}{p-1}.$$

Now $\phi(N)$ divides N implies that $p = 3$. So if N satisfies the given condition then either $N = 1$ or $N = 2^a 3^b$ where $a > 0$ and $b \geq 0$.

56. Let G be a group and $x \in G$ has order n. Show that if k is a positive integer then $\text{Order}(x^k) = \dfrac{n}{(k,n)}$

Solution:

Let $gcd(k, n) = d$, so $gcd\left(\frac{k}{d}, \frac{n}{d}\right) = 1$. Let $\frac{n}{d} = n_1$ and $\frac{k}{d} = k_1$ (so $gcd(k_1, n_1) = 1$). Let e denote the identity element of the group G. Now

$$(x^k)^{n_1} = (x^k)^{\frac{n}{d}} = (x^n)^{\frac{k}{d}} = e.$$

Now suppose that r is a natural number satisfying $(x^k)^r = e$. We only need to show that r is divisible by $n_1 = \frac{n}{d}$. Now

$$x^{kr} = e \Longrightarrow n|kr \Longrightarrow n_1|(k_1 r) \Longrightarrow n_1|r \quad \text{as } gcd(k_1, n_1) = 1.$$

57. Which of the following congruence has a solution?

(a) $x^2 \equiv -5 \pmod{997}$.

(b) $x^2 \equiv 7 \pmod{1009}$.

(c) $x^2 \equiv -163 \pmod{1999}$.

Answer: (a) No solution, (b) has a solution, (c) has a solution.

58. Without doing the actual calculations find the remainder when

$$1^{50} + 2^{50} + 3^{50} + \cdots + 99^{50} + 100^{50}$$

is divided by 101.

Solution:

Since 101 is a prime number and as $\left(\frac{a}{101}\right) \equiv a^{50} \pmod{101}$, the given sum is $\equiv 0 \pmod{101}$.

59. Each alphabet in the equation below represents a different digit. Same alphabet represent the same digit. What number does each alphabet represent? $(BE) \times (CE) = ZZZ$.

Solution: We have $BE \times CE = ZZZ$. Clearly $ZZZ = Z \times 111 = Z \times 3 \times 37$. Therefore $BE \times CE = Z \times 3 \times 37$. By unique factorization property in $\mathbb{N}$, we have that 37 divides BE or CE. So one of these should equal 37 or 74. But none of these can be 74 as then $ZZZ = 666$ which is not divisible by 4 but as $E = 4$, $BE \times CE$ is certainly divisible by 4. Hence, one of BE or CE is equal to 37 and as then $Z = 9$ the other one is $999 \div 37 = 27$. The complete solution is $27 \times 37 = 999$.

60. Each alphabet in the equation below represents a different digit. Same alphabets represent the same digit. What number does each alphabet represent?

$$
\begin{array}{r}
COFFEE \\
+COFFEE \\
+COFFEE \\
------ \\
THEOREM
\end{array}
$$

Solution: It is clear that $3 \times E$ is a two-digit number and so $E \geq 4$. The following table makes us conclude that $E = 9$ and $M = 7$:

E	$3 \times E$	Carry	$3 \times E +$ Carry
4	12	1	**13**
5	15	1	**16**
6	18	1	**19**
7	21	2	**23**
8	24	2	**26**
9	**27**	2	**29**

Since $3 \times O+$previous carry has the unit digit $E = 9$, we must have $O = 3$ (with previous carry $=$ zero, and so $F = 1$) or $O = 6$ (with previous carry $= 1$).

Now $O = 6 \Rightarrow F = 5 \Rightarrow R = 7$, which is not possible as $M = 7$. So far we have $E = 9$, $M = 7$, $O = 3$, $F = 1$, $R = 5$. Thus

$$\begin{array}{r} C31199 \\ +C31199 \\ +C31199 \\ ------- \\ TH93597 \end{array}$$

As $C \geq 4$ we have $C = 4, 6$, or 8. Since $C = 4$ or 6 would imply that $T = 1$ which is not possible as $F = 1$, so we must have $C = 8$ and the final solution is

$$\begin{array}{r} 831199 \\ +831199 \\ +831199 \\ ------- \\ 2493597 \end{array}$$

61. Each alphabet in the equation below represents a different digit. Same alphabet represent the same digit. What number does each alphabet represent? (*You may assume that* $S \neq 0$).

$$\begin{array}{r} SQUARE \\ +SQUARE \\ +SQUARE \\ +SQUARE \\ ------- \\ NUMBER \end{array}$$

Solution:

We first note that since $4E \equiv R \pmod{10}$, R must be even. We have

$$4E \equiv R \pmod{10} \Rightarrow 2E \equiv \frac{R}{2} \pmod{5} \Rightarrow E \equiv \frac{3R}{2}$$
$$\pmod{5} \Rightarrow E \equiv 4R \pmod{5}.$$

The following table makes us conclude that we must have $E = 2$, $R = 8$, and $S = 1$:

R	$E \equiv 4R \pmod{5}$	Carry from E-column	Conclusion
0	0	0	Not possible as $E \neq R$
2	8	3	$\Rightarrow E = 1$ Not possible
4	6	2	$\Rightarrow E = 8$ Not possible
6	4	1	$\Rightarrow E = 5$ Not possible
8	2	0	$\Rightarrow E = 2$ possible

Since the S-column has no carry we must have $S \leq 2$. We know that $S \neq 0$ and as $E = 2$ we must have $S = 1$. So far we have $E = 2$, $R = 8$, and $S = 1$. As carry from R-column is 3 we have $4A + 3 \equiv B \pmod{10}$, or $4A \equiv B - 3 \pmod{10}$. So B must be odd. We note that $4A \equiv B - 3 \pmod{10} \Rightarrow A \equiv 4(B-3) \pmod{5}$. The following table gives the possible values of B and A:

B	3	3	5	7	9
A	0	5	3	6	4

In what follows that we show that $B = 3$ $A = 0$ is the only possible solution.

(a) **Case I** $A = 5, B = 3$ is not possible.

In this case, the carry from the A-column is 2. So from the U-column, we have $4U + 2 \equiv M \pmod{10}$. Therefore, M must be even. Since $M \neq 2, 8$, the only possible values of M are $0, 4$, and 6.

- $M = 0 \Rightarrow U \equiv 2 \pmod{5} \Rightarrow U = 7$. As $U \neq E = 2$. Now we have $A = 5, B = 3, M = 0, U = 7, S = 1$. As the carry from the A-column is 2, the carry from the U-column is 3. Therefore we have $4Q + 3 \equiv 7$

(mod 10). This gives us $Q \equiv 1 \pmod 5$. As $Q \neq 1$, we must have $Q = 6$. But $Q = 6 \Rightarrow N = 6$, this is not possible as $Q \neq N$.

- $M = 4 \Rightarrow 4U + 2 \equiv 4 \pmod{10} \Rightarrow U = 3, 8$. As $R = 8$, $U = 8$ is not possible and so $U = 3$. $U = 3$ implies that the carry from the U-column is one. Therefore

  $4Q + 1 \equiv 3 \pmod{10} \Rightarrow Q = 3, 8$. This is not possible.
- $M = 6 \Rightarrow 4U + 2 \equiv 6 \pmod{10} \Rightarrow U \equiv 1 \pmod 5 \Rightarrow U = 1, 6$. This is not possible.

(b) **Case II** $A = 3, B = 5$ is not possible.

$A = 3, B = 5 \Longrightarrow 4U + 1 \equiv M \pmod{10} \Longrightarrow M$ is odd. Since the digits $1, 3$, and 5 have already been associated with other letters we have $M = 7$ or $M = 9$. We show that neither of these is possible.

- Suppose $M = 7$. Then $4U + 1 \equiv 7 \pmod{10} \Longrightarrow U \equiv 4 \pmod 5$. So $U = 4$ or $U = 9$.

  $U = 4 \Longrightarrow 4Q + 1 \equiv 4 \pmod{10}$ which is not possible. $U = 9 \Longrightarrow 4Q + 3 \equiv 9 \pmod{10} \Longrightarrow Q \equiv 4 \pmod 5 \Longrightarrow Q = 4$ or $Q = 9$. Since $U = 9$, $Q = 9$ is not possible. Let $Q = 4$. But $Q = 4 \Longrightarrow S = 5$ which is not possible as $B = 5$.
- Suppose that $M = 9$. Then $4U+1 \equiv 9 \pmod{10} \Longrightarrow U \equiv 2 \pmod 5 \Longrightarrow U = 2$ or $U = 7$. As $E = 2$ therefore $U = 2$ is not possible so we let $U = 7$. In this case, carry from U column is 2, so $4Q + 2 \equiv U \pmod{10} \Longrightarrow 4Q \equiv 5 \pmod{10}$ which is not possible.

(c) **Case III** $A = 6, B = 7$ is not possible.

$A = 6, B = 7 \Rightarrow 4U + 2 \equiv M \pmod{10} \Rightarrow M = 0, 4$.

- $M = 0 \Rightarrow 4U \equiv 8 \pmod{10} \Rightarrow U \equiv 2 \pmod 5 \Rightarrow U = 2, 7$. Not possible.
- $M = 4 \Rightarrow 4U \equiv 2 \pmod{10} \Rightarrow U \equiv 3 \pmod 5 \Rightarrow U = 3$.

  Now $U = 3 \Rightarrow 4Q + 1 \equiv 3 \pmod{10} \Rightarrow Q \equiv 3 \pmod 5 \Rightarrow Q = 3, 8$. Not possible.

(d) **Case IV** $A = 4, B = 9$ is not possible.

$A = 4, B = 9 \Rightarrow 4U + 1 \equiv M \pmod{10} \Rightarrow M = 3, 5, 7$.

- $M = 3 \Rightarrow 4U \equiv 2 \pmod{10} \Rightarrow U \equiv 3 \pmod 5 \Rightarrow U = 3$. Not possible.
- $M = 5 \Rightarrow 4U \equiv 4 \pmod{10} \Rightarrow U \equiv 1 \pmod 5 \Rightarrow U = 6$.

  Now $M = 5, U = 6 \Rightarrow 4Q + 2 \equiv 6 \pmod{10} \Rightarrow Q \equiv 1 \pmod 5 \Rightarrow Q = 1, 6$. Not possible.
- $M = 7 \Rightarrow 4U \equiv 6 \pmod{10} \Rightarrow U \equiv 4 \pmod 5 \Rightarrow U = 4, 9$. Not possible.

(e) **Case V** $A = 0, B = 3$ is possible.

In this case, we have $4U \equiv M \pmod{10}$. So M must be even and therefore the only possible values of M are 4 and 6.

- $M = 4 \Rightarrow U = 6 \Rightarrow Q = 1$ or $Q = 6$. Neither of these is possible.
- $M = 6 \Rightarrow U = 4$ or $U = 9$.

  $U = 4 \Rightarrow 4Q + 1 \equiv 4 \pmod{10}$. This is not possible.

  $U = 9 \Rightarrow Q = 4$ and so $N = 5$ and we have the following solution to our problem:

A	B	E	M	N	Q	R	S	U
0	3	2	6	5	4	8	1	9

62. Find all integers N such that $\phi(N) = 144$.

Solution:

The only primes which can appear in the prime factorization of such N are $2, 3, 5, 7, 13, 19, 37,$ and 73. The prime 17 is excluded because $\phi(n) = 9$ has no solutions. A case by case discussion like the one discussed in case of finding the solutions for $\phi(n) = 12$ yields that there are 21 numbers N such that $\phi(N) = 144$. These 21 numbers are

185	219	273	285	292	296	304
315	364	370	380	432	438	444
456	468	504	540	546	570	630

63. Find the least positive integer k such that $3^k \equiv 1 \pmod{59}$

Answer:

29

64. Which of the following congruence have solutions? Can you find these solutions?

(a) $x^2 \equiv 30 \pmod{1973}$

(b) $x^2 \equiv 28 \pmod{37}$

Answer:

(a) No solutions.

(b) $x \equiv 18, 19 \pmod{37}$.

65. Solve the congruence: $X^2 \equiv 94 \pmod{13^3}$

Answer: $X \equiv \pm 225 \pmod{13^3}$

66. Solve the congruence: $2X^2 + 5X - 9 \equiv 0 \pmod{101}$

Answer: $X \equiv 19, 29 \pmod{101}$

67. Find the smallest positive X satisfying the congruence $X^2 \equiv 2 \pmod{26711}$

Solution:

Since $26711 \equiv 3 \pmod 4$, $X \equiv 2^{\frac{26712}{4}} = 2^{6678} \pmod{26711}$ We can calculate $2^{6678} \pmod{26711}$ as follows.

We first note that in binary $6678 = 1101000010110$.

y	d	$y^2 * 2^d \pmod{26711}$
1	1	2
2	1	8
8	0	64
64	1	8192
8192	0	10832
10832	0	17512
17512	0	1153
1153	0	$20570 \equiv -6141$
6141	1	$18609 \equiv 8102$
8102	0	13477
13477	1	16169
16169	1	5297
5297	0	11659

68. Prove that $(x, y) = (1, 0)$ is the only solution of the Diophantine equation $y^2+1 = x^3$.

Proof:

First, we notice that x must be odd and y must be even. For if x were even then x^3 would be a multiple of 8 and $y^2 + 1$ cannot be a multiple of 8.

We can write the above equation as $(y + i)(y - i) = x^3$. We claim that in $\mathbb{Z}[i]$ $GCD(y+i, y-i) = 1$. Let α be a common divisor of $y+i$ and $y-i$. We show that

$N(\alpha) = 1$ and so it must be a unit in $\mathbb{Z}[i]$. The fact that α divides $y+i$ (in $\mathbb{Z}[i]$) implies that $N(\alpha)$ divides y^2+1 in $\mathbb{Z}$. As y is even y^2+1 is odd and so $N(\alpha)$ is odd. As α is a divisor of both $y+i$ and $y-i$ it divides $2i$. This implies that $N(\alpha)$ divide 4 in $\mathbb{Z}$. But 1 is the only odd divisor of 4. Hence $N(\alpha)=1$ and therefore α is a unit. This shows that $GCD(y+i, y-i) = 1$. Therefore, both $y+i$ and $y-i$ are cubes of some Gaussian integers (note that this is so as every unit in $\mathbb{Z}[i]$ is a cube of some unit). Suppose that $y+i = (a+bi)^3$ where $a, b \in \mathbb{Z}$. Then we have

$$y + i = a^3 - 3ab^2 + i(3a^2b - b^3).$$

Therefore $3a^2b - b^3 = b(3a^2 - b^2) = 1$ which is possible only if $b = -1$ and $a = 0$. This in turn implies that $y = 0$ and $x = 1$.

69. Prove that

$$(1 + x + x^2 + x^3 + \cdots + x^n)^2 - x^n = \prod_{\substack{k|n \\ k>1}} \Phi_k(x) \prod_{\substack{l|(n+2) \\ l>1}} \Phi_l(x)$$

Proof:

Let $A = 1+x+x^2+\cdots+x^n$ and $B = (1+x+x^2+x^3+\cdots+x^n)^2 - x^n = A^2 - x^n$. We note that $A^2 + A(x^{n+1} - x^n) - x^{2n+1} = (A - x^n)(A + x^{n+1})$. We have

$$\begin{aligned} B &= A^2 - x^n \\ &= A^2 + x^n(x^{n+1} - 1) - x^{2n+1} \\ &= A^2 + x^n(x-1)(1 + x + x^2 + \cdots + x^n) - x^{2n+1} \\ &= A^2 + A(x^{n+1} - x^n) - x^{2n+1} \\ &= (A - x^n)(A + x^{n+1}) \\ &= (1 + x + x^2 + \cdots + x^{n-1})(1 + x + x^2 + \cdots + x^n + x^{n+1}) \\ &= \frac{x^n - 1}{x - 1}\frac{x^{n+2} - 1}{x - 1}. \end{aligned}$$

Hence

$$B = \prod_{\substack{k|n \\ k>1}} \Phi_k(x) \prod_{\substack{l|(n+2) \\ l>1}} \Phi_l(x).$$

70. Let $\omega = \dfrac{-1+\sqrt{-3}}{2}$ and let $\pi = a + b\omega \in \mathbb{Z}[\omega]$ be a prime in $\mathbb{Z}[\omega]$ such that $N(\pi) = a^2 - ab + ab = p \equiv 1 \pmod 3$ is a prime in $\mathbb{Z}$. Show that among six

associates of π there is just one which if written as $A + B\omega$ is such that $A \equiv 2 \pmod 3$ and $B \equiv 0 \pmod 3$.

Solution:

As $\langle -\omega \rangle = \{1, -\omega, \omega^2, -1, \omega, -\omega^2\}$ the six associates of $\pi = a + b\omega$ are obtained by multiplying π by six elements of $\langle -\omega \rangle$. These six associates are

$$a + b\omega, b + (b - a)\omega, (b - a) - a\omega, -1 - b\omega, -b + (a - b)\omega, (a - b) + a\omega.$$

- **Case 1** $\quad a \equiv 0 \pmod 3$ and $b \not\equiv 0 \pmod 3$.

  In this case if $b \equiv 2 \pmod 3$ then $b - a \equiv 2 \pmod 3$ and so $(b - a) - a\omega$ has the desired form.

  If $b \equiv 1 \pmod 3$ then $a - b \equiv 2 \pmod 3$ and so $(a - b) + a\omega$ has the desired form.
- **Case 2** $\quad b \equiv 0 \pmod 3$ and $a \not\equiv 0 \pmod 3$.

  In this case $a + b\omega$ or $-a - b\omega$ has the desired form.
- **Case 3** $\quad ab \not\equiv 0 \pmod 3$.

  If $a \equiv 1 \pmod 3$ and $b \equiv 2 \pmod 3$ then $a^2 - ab + b^2 \not\equiv 0 \pmod 3$.

  If $a \equiv 2 \pmod 3$ and $b \equiv 1 \pmod 3$ then $a^2 - ab + b^2 \not\equiv 0 \pmod 3$.

  If $a \equiv b \equiv 1 \pmod 3$ then $-b + (a - b)\omega$ has the desired form.

  If $a \equiv b \equiv 2 \pmod 3$ then $b + (b - a)\omega$ has the desired form.

71. Let $p \in \mathbb{Z}$ be a prime such that $p \equiv 1 \pmod 3$. Show that there exist integers $A, B \in \mathbb{Z}$ such that $A^2 + 27B^2 = 4p$.

Solution:

We know that there exist $x, y \in \mathbb{Z}$ such that $x^2 + 3y^2 = p$. If y is divisible by 3 then we have nothing to prove. Suppose $3 \nmid y$. Two cases arise.

- $x + y \equiv 0 \pmod 3$

  In this case, we have

  $$(x - 3y)^2 + 3(x + y)^2 = 4(x^2 + 3y^2) = 4p.$$
- $x - y \equiv 0 \pmod 3$

  In this case, we have

  $$(x + 3y)^2 + 3(x - y)^2 = 4(x^2 + 3y^2) = 4p.$$

72. Let $p \in \mathbb{Z}$ be a prime, $p \equiv 1 \pmod 3$. Show that there exist $a, b \in \mathbb{Z}$ such that $p = a^2 - ab + b^2$.

Solution:

Using Thue's theorem, we can find $x, y \in \mathbb{Z}$ such that $x^2+3y^2 = p$. Let $a = x+y$ and $b = 2y$. Then we have

$$a^2 - ab + b^2 = (x+y)^2 - 2y(x+y) + (2y)^2 = x^2 + 3y^2 = p.$$

73. Let $A = [a_{i,j}]$ and $B = [b_{j,k}]$ be $n \times n$ matrices, where

$$a_{i,j} = \begin{cases} 1 \text{ if } j|i \\ 0 \text{ otherwise} \end{cases} \qquad b_{j,k} = \begin{cases} \mu\left(\frac{j}{k}\right) \text{ if } k|j \\ 0 \text{ otherwise} \end{cases}$$

where μ is Mobius μ-function, then $A \cdot B = I_n$, where I_n is $n \times n$ identity matrix.

Solution:

Let $A \cdot B = C = [c_{i,k}]$ so we have

$$[a_{i,j}] \cdot [b_{j,k}] = [c_{i,k}], \qquad 1 \le i, j, k \le n.$$

So $c_{i,k} = \sum_{j=1}^{n} a_{i,j} b_{j,k}$. Since $a_{i,j} \neq 0 \Longrightarrow j|i$ we need to sum for $j = 1$ to $j = i$ only. Therefore $c_{i,k} = \sum_{j=1}^{i} a_{i,j} b_{j,k}$. Now

$$a_{i,j} b_{j,k} \neq 0 \Longrightarrow j|i \text{ and } k|j \Longrightarrow k|i$$

Assume that $a_{i,j} b_{j,k} \neq 0$. Then $i = rk$, $j = sk$ for some $r, s \in \mathbb{N}$. We show that $c_{i,k} \neq 0 \Longrightarrow i = k$ and this will imply that $C = I_n$. We have

$$\begin{aligned} c_{i,k} &= \textstyle\sum_{j=1}^{n} a_{i,j} b_{j,k} = \sum_{j=1}^{i} a_{i,j} b_{j,k} = \sum_{j|i} a_{i,j} b_{j,k} = \sum_{j|i} b_{sk,k} \\ &= \textstyle\sum_{sk|rk} b_{sk,k} = \sum_{s|r} b_{sk,k} = \sum_{s|r} \mu\left(\frac{sk}{k}\right) = \sum_{s|r} \mu(s). \end{aligned}$$

As proved earlier $\sum_{s|r} \mu(s) = 1$ if $r = 1$ and zero otherwise. Therefore $c_{i,k} \neq 0 \Longrightarrow r = 1$, $i = rk = k$ and so $c_{i,k} = c_{i,i} = 1$ for $i \le i \le n$. Therefore $C = A \cdot B$ is an $n \times n$ identity matrix.

74. Let $p \equiv 2 \pmod 3$ be a prime number. Let $a, b \in \mathbb{Z}$ be such that p divides a^2+ab+b^2. Prove that p divides both a and b.

First Solution:

We note that $p \equiv 2 \pmod 3 \implies \left(\frac{-3}{p}\right) = -1$. Now $p|(a^2+ab+b^2)$ implies that p divides $4(a^2+ab+b^2) = (2a+b)^2+3b^2$. So if p does not divide b then $\left(\frac{-3}{p}\right) = 1$. Hence p divides b and also a.

Second Solution:

We know that a rational prime $p \equiv 2 \pmod 3$ is a prime in $\mathbb{Z}[\omega]$. Now as $a^2+ab+b^2 = (a+b\omega)(a+b\omega^2)$ and since p is a prime in $\mathbb{Z}[\omega]$, p divides $a+b\omega$ or $a+b\omega^2$. So either $\frac{a+b\omega}{p}$ or $\frac{a+b\omega^2}{p}$ lie in $\mathbb{Z}[\omega]$. This can happen only if p divides both a and b.

75. Let p be prime, $p \equiv 3 \pmod 4$. Prove that $a^2+b^2 \equiv 0 \pmod p$ imply that $p|a$ and $p|b$.

Proof:

If p does not divide a then $ax \equiv 1 \pmod p$ for some integer x. Multiplying the congruence $a^2+b^2 \equiv 0 \pmod p$ by x^2 we get $(ax)^2+b^2x^2 \equiv 0 \pmod p$ or $1+(bx)^2 \equiv 0 \pmod p$ which implies that $\left(\frac{-1}{p}\right) = 1$. This is not true.

76. Prove that a prime $p > 3$ can be represented as $p = x^2+xy+y^2$ if $p \equiv 1, 7 \pmod{12}$. A prime $p \equiv 5, 11 \pmod{12}$ cannot be so represented.

Solution:

First, we show that a prime $p \equiv 5, 11 \pmod{12}$ cannot be represented as x^2+xy+y^2. For primes $p \equiv 5, 11 \pmod{12}$, we have

$$x^2+xy+y^2 = p \Longrightarrow (2x+y)^2+3y^2 = 4p \Longrightarrow \left(\frac{-3}{p}\right) = 1.$$

But $1 = \left(\frac{-3}{p}\right) = \left(\frac{p}{3}\right) = \left(\frac{2}{p}\right) = -1.$

Now let $p \equiv 1, 7 \pmod{12}$. In this case as $\left(\frac{-3}{p}\right) = 1$, we can find $a \in \mathbb{Z}$ such that $a^2 \equiv -3 \pmod p$. By Thue's theorem, there exist $x, y \in \mathbb{Z}$ such that $a \equiv \pm\frac{x}{y} \pmod p$, $x, y < \sqrt{p}$, or $x^2+3y^2 \equiv 0 \pmod p$ $x, y < \sqrt{p}$. Therefore $x^2+3y^2 = mp$, where $m = 1, 2$, or 3.

Case 1 $\quad m = 1$.

In this case, we have $x^2+3y^2 = p$ which implies that $(x-y)^2+(x-y)2y+(2y)^2 = p$.

Case 2 $m = 2$.

In this case, $x^2 + 3y^2 = 2p$ and so x and y have same parity and therefore $x^2 + 3y^2 \equiv 0 \pmod 4$, but $2p$ is not divisible by 4.

Case 3 $m = 3$.

In this case, $x^2 + 3y^2 = 3p$ and so x is divisible by 3. We let $x = 3x_1$ and so $3x_1^2 + y^2 = p$ which reduces to case 1.

77. Find the smallest positive integer giving remainders $1, 2, 3, 4$, and 5 when divided by $3, 5, 7, 9$, and 11, respectively.

Since $x \equiv 7 \pmod 9 \implies x \equiv 1 \pmod 3$, we can ignore the condition $x \equiv 1 \pmod 3$. Now

$x \equiv 4 \pmod 9$ and $x \equiv 5 \pmod{11} \implies x \equiv 49 \pmod{9 \times 11}$

$x \equiv 49 \pmod{99}$ and $x \equiv 3 \pmod 7 \implies x \equiv 346 \pmod{7 \times 9 \times 11}$

$x \equiv 346 \pmod{7 \times 99}$ and $x \equiv 2 \pmod 5 \implies x \equiv \mathbf{1732} \pmod{5 \times 7 \times 9 \times 11}$.

78. If q and $p = 4q + 1$ are primes then prove that 2 is a primitive root modulo p.

Solution:

Since the divisors of $p - 1 = 4q$ are $1, 2, 4, q, 2q, 4q$, it is enough to show that $2^{2q} \not\equiv 1 \pmod p$. As q is an odd prime $p = 4q + 1 \equiv 5 \pmod 8$. As

$$-1 = \left(\frac{2}{p}\right) \equiv 2^{\frac{p-1}{2}} = 2^{2q} \pmod p$$

the order of 2 must be $4q = p - 1$.

79. Let $p \equiv 11 \pmod{12}$ be a prime number. Prove that there exist $x, y, a, b, \in \mathbb{Z}$ such that $2p = 6x^2 - 2y^2 = a^2 - 3b^2$.

Solution:

We first observe that $n = 3x^2 - y^2 \implies 2n = (3x + y)^2 - 3(x + y)^2$, and

$$n \text{ odd}, \quad 2n = x^2 - 3y^2 \implies n = 3\left(\frac{x-y}{2}\right)^2 - \left(\frac{x-3y}{2}\right)^2.$$

So all we need to show is that every prime $p \equiv 11 \pmod{12}$ can be expressed as $x^2 - 3y^2$, $x, y \in \mathbb{N}$. Now as

$$p \equiv 11 \pmod{12} \implies \left(\frac{3}{p}\right) = -\left(\frac{p}{3}\right) = -\left(\frac{2}{3}\right) = 1$$

$a^2 \equiv 3 \pmod 3$ for some integer a. By Thue's theorem, we can find $x, y \in \mathbb{N}$, $x, y < \sqrt{p}$ and $\pm\frac{y}{x} \equiv a \pmod p$. This gives $y^2 - 3x^2 \equiv 0 \pmod p$. Since $x, y < \sqrt{p}$, $y^2 > 3x^2$ and $y^2 < p$ would imply that $y^2 - 3x^2 = 0$ which is not possible. So $3x^2 - y^2 > 0$, $3x^2 - y^2 \equiv 0 \pmod p$ and $3x^2 - y^2 = p$ or $3x^2 - y^2 = 2p$. $3x^2 - y^2 = p$ implies that x, y have the same parity. As x, y both cannot be even they must be both odd. Then $3x^2 - y^2 \equiv 2 \pmod 8$ but $2p \equiv 6 \pmod 8$.

80. Let $q \equiv 1 \pmod 4$ be a prime such that $p = 2q + 1$ is also a prime. Prove that 2 is a primitive root modulo p.

Solution:

It is sufficient to show that the order of 2 modulo p is not q. Now

$$q \equiv 1 \pmod 4, \text{ and } p = 2q + 1 \implies p \equiv 3 \pmod 8.$$

Therefore

$$-1 = \left(\frac{2}{p}\right) \equiv 2^{\frac{p-1}{2}} = 2^q \pmod p$$

and so the order of 2 modulo p is not q.

81. Prove that a prime $p \equiv 3 \pmod 8$ can be expressed as a sum of three squares.

Hint:

If p is a prime $\equiv 3 \pmod 8$ then $\left(\frac{-2}{p}\right) = 1$. Using Thue's theorem show that a prime can be written as $x^2 + 2y^2 = x^2 + y^2 + y^2$.

Note:

Most (but not all) of the primes $p \equiv 3 \pmod 8$ can be expressed as a sum of three different squares. For example, $59 = 49 + 9 + 1$, $107 = 81 + 25 + 1$, and $139 = 81 + 49 + 9$. It seems that every prime $p \equiv 3 \pmod 8$, $p > 163$ can be expressed as a sum of three *different* squares.

82. Solve the congruence: $X^2 \equiv 2 \pmod{61441}$.

Solution:

Following the notation used in the method of computing the square root of the primes $p \equiv 1 \pmod 4$ discussed in the text we have

$a = 2, \qquad h = 11, \qquad e_1 = 30720, \qquad e_2 = 61440.$

e_1	e_2	a^{e_1}	h^{e_2}	$a^{e_1} \cdot h^{e_2}$	E_2
30720	61440	1	1	1	61440
15360	30720	1	-1	-1	61440
7680	30720	-1	-1	1	30720
3840	15360	28946	28946	-1	46080
1920	23040	44853	5167	-1	53760
960	26880	19848	9587	-1	57600
480	28800	34468	46423	1	28800
240	14400	23248	30739	1	14400
120	7200	39944	59769	-1	37920
60	18960	25085	22083	-1	49680
30	24840	60349	31677	-1	55560
15	27780	32768	7682	-1	58500
8	29250	256	1285	Answer: 21755	

Therefore $21755^2 \equiv 2 \pmod{61441}$.

83. Find all primes p, $2 < p < 500$ satisfying the following conditions:

$$\left(\frac{2}{p}\right) = \left(\frac{3}{p}\right) = \left(\frac{5}{p}\right) = 1.$$

List your answers in ascending order.

Solution:

We first note that

$$\left(\frac{2}{p}\right) = 1 \Longrightarrow p \equiv 1, 7 \pmod{8}$$

$$\left(\frac{3}{p}\right) = 1 \Longrightarrow p \equiv 1, 11 \pmod{12}$$

$$\left(\frac{5}{p}\right) = 1 \Longrightarrow p \equiv 1, 4 \pmod{5}$$

Now

$p \equiv 1 \pmod 8, \quad p \equiv 1 \pmod 5 \Longrightarrow p \equiv 1 \pmod{40}$

$p \equiv 1 \pmod 8, \quad p \equiv 4 \pmod 5 \Longrightarrow p \equiv 9 \pmod{40}$

$p \equiv 7 \pmod 8, \quad p \equiv 1 \pmod 5 \Longrightarrow p \equiv 31 \pmod{40}$

$p \equiv 7 \pmod 8, \quad p \equiv 4 \pmod 5 \Longrightarrow p \equiv 39 \pmod{40}$

Combining these with $p \equiv 1, 11 \pmod{12}$, we get the following possibilities for primes $2 < p < 500$:

(a) $p \equiv 1 \pmod{40}, \quad p \equiv 1 \pmod{12} \Longrightarrow p \equiv 1 \pmod{120}$. So $p = 241$.

(b) $p \equiv 9 \pmod{40}, \quad p \equiv 1 \pmod{12} \Longrightarrow p \equiv 49 \pmod{120}$. So $p = 409$.

(c) $p \equiv 31 \pmod{40}, \quad p \equiv 11 \pmod{12} \Longrightarrow p \equiv 71 \pmod{120}$.
So $p = 71, 191, 311, 431$.

(d) $p \equiv 39 \pmod{40}, \quad p \equiv 11 \pmod{12} \Longrightarrow p \equiv 119 \pmod{120}$.
So $p = 239, 359, 479$.

So the primes in ascending order satisfying the given conditions are

$$71, 191, 239, 241, 311, 359, 409, 431, 479.$$

84. Find **all** the solutions of the congruence $x^2 \equiv 2 \pmod{19303}$.

Solution:

We note that $19303 = 97 \times 199$. So first we have to solve two congruence $x^2 \equiv 2 \pmod{97}$ and $x^2 \equiv 2 \pmod{199}$. We have

$x^2 \equiv 2 \pmod{97} \Longrightarrow x \equiv 14, 83 \pmod{97}$

$x^2 \equiv 2 \pmod{199} \Longrightarrow x \equiv 20, 179 \pmod{199}$.

Now

$x \equiv 14 \pmod{97}$ and $x \equiv 20 \pmod{199} \Longrightarrow x \equiv 11363 \pmod{19303}$

$x \equiv 83 \pmod{97}$ and $x \equiv 179 \pmod{199} \Longrightarrow x \equiv 7940 \pmod{19303}$

$x \equiv 14 \pmod{97}$ and $x \equiv 179 \pmod{199} \Longrightarrow x \equiv 12915 \pmod{19303}$

$x \equiv 83 \pmod{97}$ and $x \equiv 20 \pmod{199} \Longrightarrow x \equiv 6388 \pmod{19303}$.

Therefore, the four solutions of the given congruence are $x \pm 7940 \pmod{19303}$ and $x \equiv \pm 6388 \pmod{19303}$.

85. Does the congruence $x^2 \equiv 4042 \pmod{4297}$ have a solution? Equivalently calculate the value of Legendre symbol $\left(\dfrac{4042}{4297}\right)$.

Solution:

We have

$$\begin{aligned}
\left(\frac{4042}{4297}\right) &= \left(\frac{2021}{4297}\right), && \text{Since } \left(\frac{2}{4297}\right) = 1. \\
&= \left(\frac{4297}{2021}\right), && \text{By Reciprocity.} \\
&= \left(\frac{255}{2021}\right), && \text{Since } 4297 \equiv 255 \pmod{2021}. \\
&= \left(\frac{2021}{255}\right), && \text{By Reciprocity.} \\
&= \left(\frac{236}{255}\right), && \text{Since } 2021 \equiv 236 \pmod{255}. \\
&= \left(\frac{59}{255}\right), && \text{Since } 236 = 4 \times 59. \\
&= -\left(\frac{255}{59}\right), && \text{By Reciprocity.} \\
&= -\left(\frac{19}{59}\right), && \text{Since } 255 \equiv \pmod{59}. \\
&= \left(\frac{59}{19}\right), && \text{By Reciprocity.} \\
&= \left(\frac{2}{19}\right), && \text{since } 59 \equiv 2 \pmod{19}. \\
&= -1, && \text{Since } 19 \equiv 3 \pmod{8}.
\end{aligned}$$

Therefore, the given congruence has no solution.

86. Find the smallest positive integer N consisting only of digits 3 and 7 (and both these digits do appear in the number) such that both this number and the sum of its digits are divisible by both 3 and 7.

Solution:

Let N denote the smallest number with stated properties. Let x and y denote the number of threes and number of sevens in decimal representation of N. Then $3x + 7y \equiv 0 \pmod{21}$. The smallest positive solution for this congruence is $x = 7$ and $y = 3$. We find a smallest 10-digit number consisting of 7 threes and 3 sevens which is divisible by 21. We show that the smallest such number is 3333377733.

Any 10-digit number consisting of 7 threes and 3 sevens is obviously divisible by 3. We need to find the positions of 7 threes in such a number N so that it is divisible by 7. Consider the following table (note that $10 \equiv 3 \pmod 7$):

i	10	9	8	7	6	5	4	3	2	1
$10^i \equiv 3^i \pmod 7$	4	6	2	3	1	5	4	6	2	3

As we want N to be the smallest number with the desired properties, we need to have the digit 3 appear on as many places on the left of the number as possible. Not all the seven leading digits can be 3 as $\sum_{i=1}^{i=7} 3^{3+i} \equiv 25 \equiv 4 \pmod 7$. If we let the first six leading digits to be 3 then $\sum_{i=2}^{i=7} 3^{3+i} \equiv 21 \equiv 0 \pmod 7$ and this means that in whatever position we keep another digit 3 the number N will not be divisible by 7. We try to put the digit 3 in the five leading positions. This gives $\sum_{i=3}^{i=7} 3^{3+i} \equiv 16 \equiv 2 \pmod 7$. We need to find two positions among the last five to place the digit 3 at these positions with the conditions that the contribution to number N from these two placement of the digit 3 is congruent to $5 = 7 - 2 \pmod 7$. We find from the second row of the above table that the only possibility for that to happen is to put one digit 3 at unit's position and other at ten's position corresponding to $i = 1$ and $i = 2$ in our table. If we do that then we will have

$$N = 3^{10} + 3^9 + 3^8 + 3^7 + 3^6 + 7^5 + 7^4 + 7^3 + 3^2 + 3^1 \equiv 4+6+2+3+1+2+3 = 21 \equiv 0 \pmod 7.$$

Therefore, the desired number is **3333377733**.

87. Find a necessary condition for a prime p to be simultaneously expressible as x^2+y^2, u^2+2v^2, r^2+3s^2, & m^2+5n^2. Find the least prime p satisfying this condition and express it in each of the given forms.

Solution:

$$\begin{aligned}
p = x^2 + y^2 &\implies p \equiv 1 \pmod 4 \\
p = u^2 + 2v^2 &\implies p \equiv 1, 3 \pmod 8 \\
p = r^2 + 3s^2 &\implies p \equiv 1, 7 \pmod{12} \\
p = m^2 + 5n^2 &\implies p \equiv 1, 9 \pmod{20}
\end{aligned}$$

This means that we only have to deal with the following two cases:

- $p \equiv 1 \pmod 8, \equiv 1 \pmod{12}, p \equiv 1 \pmod{20} \implies p \equiv 1 \pmod{120}$. First two primes satisfying this condition are $p = 241$ and $p = 601$.

- $p \equiv 1 \pmod 8, \equiv 1 \pmod{12}, p \equiv 9 \pmod{20} \Longrightarrow p \equiv 49 \pmod{120}$. First two primes satisfying this condition are $p = 409$ and $p = 769$.

We have

$$\begin{aligned} 241 &= 4^2 + 15^2 = 13^2 + 2 \cdot 6^2 = 7^2 + 3 \cdot 8^2 = 14^2 + 5 \cdot 3^2 \\ 601 &= 5^2 + 24^2 = 23^2 + 2 \cdot 6^2 = 13^2 + 3 \cdot 12^2 = 14^2 + 5 \cdot 9^2 \\ 409 &= 20^2 + 3^2 = 11^2 + 2 \cdot 12^2 = 19^2 + 3 \cdot 4^2 = 2^2 + 5 \cdot 9^2 \\ 769 &= 25^2 + 12^2 = 11^2 + 2 \cdot 18^2 = 1^2 + 3 \cdot 16^2 = 7^2 + 5 \cdot 12^2. \end{aligned}$$

88. Prove that if an integer k can be represented as $k = x^2 - 2y^2$ for some integers x and y, then it can be represented in this form in infinitely many different ways.

Hint:

It follows from the observation that Pell's equation $x^2 - 2y^2 = 1$ has infinitely many solutions.

89. Suppose that $p > 2$ is a prime and x, y are integers such that $2x^2 + 2xy + 3y^2 = p$. Show that $p \equiv 3, 7 \pmod{20}$. Express the primes 103 and 107 in the above form. That is find integers x, y, a, b satisfying

$$2x^2 + 2xy + 3y^2 = 103$$

$$2a^2 + 2ab + 3b^2 = 107$$

Solution:

$$2x^2 + 2xy + 3y^2 = p \Longrightarrow (2x + y)^2 + 5y^2 = 2p \Longrightarrow \left(\frac{-5}{p}\right) = 1 \Longrightarrow p \equiv 1, 9, 3, 7 \pmod{20}.$$

Since $(2x + y)^2 + 5y^2 = 2p \Longrightarrow y \equiv 1 \pmod 2 \Longrightarrow p \equiv 3 \pmod 4$, $p \equiv 1, 9 \pmod{20}$ is not possible. We conclude that $p \equiv 3, 7 \pmod{20}$.

If $(x, y) = (2, 5)$ then $2x^2 + 2xy + 3y^2 = 103$.

If $(a, b) = (5, 3)$ then $2a^2 + +2ab + 3b^2 = 107$.

90. Let $\alpha = exp(2\pi i/13) = \cos(2\pi/13) + i \sin(2\pi/13)$. Note that $\alpha^{13} = 1$ and therefore $1 + \alpha + \alpha^2 + \cdots + \alpha^{12} = 0$. Define

$$\begin{aligned} \beta_1 &= \alpha + \alpha^3 + \alpha^9 \\ \beta_2 &= \alpha^2 + \alpha^6 + \alpha^5 \\ \beta_3 &= \alpha^4 + \alpha^{12} + \alpha^{10} \\ \beta_4 &= \alpha^8 + \alpha^{11} + \alpha^7 \end{aligned}$$

Let

$$f(X) = (X - \beta_1)(X - \beta_2)(X - \beta_3)(X - \beta_4)$$

Find the polynomial $f(X)$.

Answer: $f(X) = X^4 + X^3 + 2X^2 - 4X + 3$.

91. Let $p > 2$ be a prime. Prove that

$$\left(\frac{7}{p}\right) = 1 \text{ if and only if } p \equiv 1, 3, 9, 19, 25, 27 \pmod{28}.$$

Proof:

$$p \equiv 1 \pmod 4 \text{ and } \left(\frac{7}{p}\right) = 1 \Longrightarrow p \equiv 1, 9, 25 \pmod{28}.$$

$$p \equiv 3 \pmod 4 \text{ and } \left(\frac{7}{p}\right) = 1 \Longrightarrow p \equiv 3, 19, 27 \pmod{28}.$$

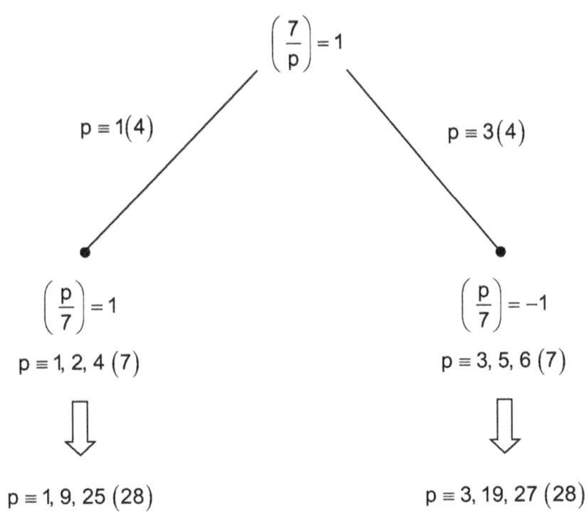

92. Let $p > 3$ be a prime. Prove that

$$\left(\frac{14}{p}\right) = 1 \text{ if and only if } p \equiv 1, 5, 9, 11, 13, 25, 29, 31, 43, 47, 51, 55 \pmod{56}.$$

Proof:

$$p \equiv \pm 1 \pmod 8 \text{ and } \left(\frac{14}{p}\right) = 1 \Longrightarrow p \equiv 1, 9, 25, 31, 47, 55 \pmod{56}.$$

$$p \equiv \pm 3 \pmod 8 \text{ and } \left(\frac{14}{p}\right) = 1 \Longrightarrow p \equiv 5, 11, 13, 29, 43, 51 \pmod{56}.$$

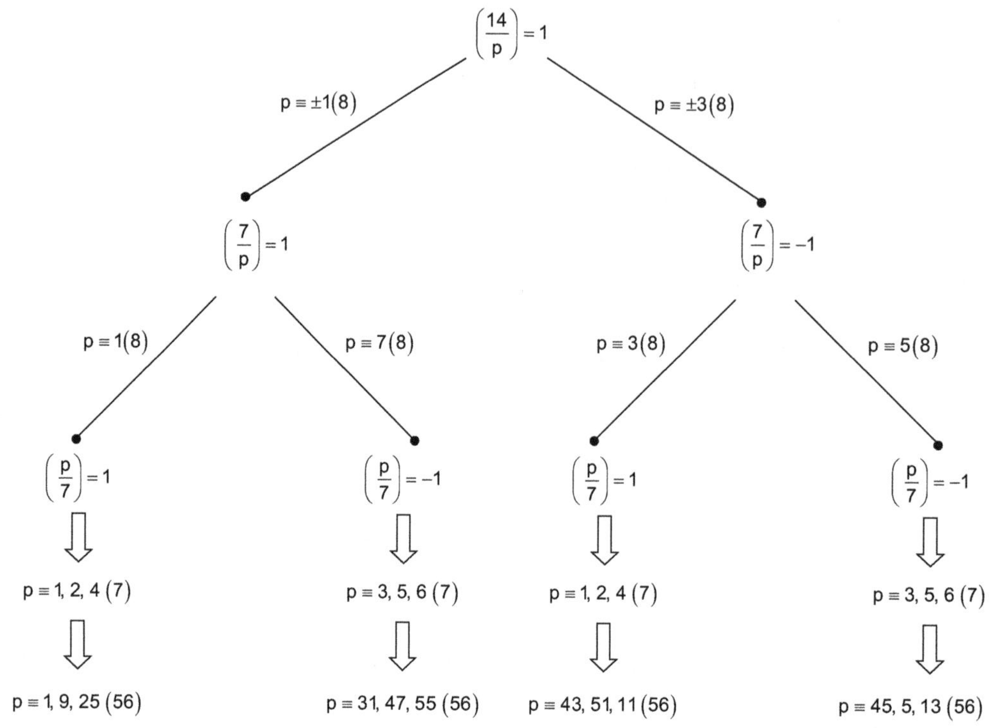

93. Solve the congruence $1000x \equiv 1 \pmod{101^3}$.

Solution:

We first solve the congruence $1000x \equiv 1 \pmod{101}$. Since

$$10 \cdot 1000 = 9999 + 1 = 99 \cdot 101 + 1 \equiv 1 \pmod{101}$$

we have $x = 10 \pmod{101}$. Let $x = 10 + 101y$. We now consider the congruence $1000x \equiv 1 \pmod{101^2}$. Substituting $x = 10 + 101y$ we get $1000(10 + 101y) \equiv 1 \pmod{101^2}$. This gives $y \equiv 20 \pmod{101}$. Let $y = 20 + 101z$. Substituting this value of y in $x = 10 + 101y$, we get

$$x = 10 + 101(20 + 101z) = 10 + 20 \cdot 101 + 101^2 z.$$

Substituting this value in the original equation, namely $1000x \equiv 1 \pmod{101^3}$ we get

$$1000 \cdot 10 + 1000 \cdot 20 \cdot 101 + 1000 \cdot 101^2 z \equiv 1 \pmod{101^3}.$$

This gives $z \equiv 30 \pmod{101}$. So let $z = 30 + 101w$. Substituting for z in $x = 10 + 20 \cdot 101 + 101^2 z$, we get

$$x = 10 + 20 \cdot 101 + 101^2(30 + 101w) = 308060 + 101^3 \cdot w \equiv 308060 \pmod{101^3}.$$

Using Hensel's lemma, this problem can be solved as follows:

We have $f(x) = 1000x - 1$, $f'(x) = 1000$, $1000^{-1} \equiv 10 \pmod{101}$, $x_1 = 10$.

i	x_i	$f(x_i) \pmod{11^{i+1}}$	$x_{i+1} = x_i - 10 \cdot f(x_i) \pmod{101^{i+1}}$
1	10	9999	$10 - 10 \cdot 9999 = -99980 \equiv 2030$
2	2030	999698	$2030 - 10 \cdot 999698 = -9994950 \equiv \mathbf{308060}$

94. Describe *all* two n-digit positive integers a and b such that the product $a \cdot b$ divides $10^n a + b$.

Solution:

ab divides $10^n a + b$ implies that a divides b and b divides $10^n a$. Now

$a|b$ and b divides $10^n a$ implies that $\frac{b}{a}$ divides 10^n. Since both a and b are n-digit numbers $\frac{b}{a} < 10$ and so $\frac{b}{a} \in \{1, 2, 4, 5, 8\}$. We deal with each of these cases separately.

- **Case 1** $\frac{b}{a} = 1$.

  In this case, $a = b$ and so a^2 divides $10^n a + a$ and so a divides $10^n + 1$. So either $a = 1$ or $10^n + 1$ has a n-digit divisor. If $a = 1$, then $a = b = 1$ and since a and b are n-digit number $n = 1$ and so $10^n a + b = 11$. If $a > 1$ then $n > 1$ (for $n = 1$ would imply that ab divides 11) and $10^n + 1$ has a as an n-digit divisor and we have

$$\frac{10^n + 1}{a} \leq \frac{10^n + 1}{10^{n-1}} = 10 + \frac{1}{10^{n-1}}.$$

  Since $\frac{10^n + 1}{a}$ is odd and $10^n + 1$ is not divisible by 3, 5, 9 we must have $\frac{10^n + 1}{a} = 7$, and so $a = \frac{10^n + 1}{7}$. Now

$$10^n + 1 \equiv 0 \pmod{7} \Longrightarrow 3^n \equiv -1 \pmod{7} \Longrightarrow 3^n \equiv 3^3 \pmod{7} \Longrightarrow 3^{n-3} \equiv 1 \pmod{7}.$$

Therefore $n \equiv 3 \pmod 6$. This gives us the following first three values of n, namely $n = 3, 9, 15$:

n	$a = b = \frac{10^n+1}{7}$	$10^n a + b$
3	143	143143
9	142, 857, 143	142, 857, 143, 142, 857, 143
15	142, 857, 142, 857, 143	142, 857, 142, 857, 143, 142, 857, 142, 857, 143

- **Case 2** $\frac{b}{a} = 2$.

In this case we have $b = 2a$. So if $n = 1$ then we have the following possibilities, for (a, b), $(1, 2)$, $(2, 4)$, $(3, 6)$, and $(4, 8)$. The last one is not possible since 32 does not divide $10^1 \cdot 4 + 8 = 48$. The following table gives the four numbers satisfying the given condition for $n = 1$:

a	1	2	3
b	2	4	6
n	12	24	36

Now suppose that $n > 1$. In this case a must be odd because

$$a \cdot b|(10^n a + b) \Longrightarrow 2a^2|(10^n a + 2a) \Longrightarrow 2a|(10^n + 2) \Longrightarrow a \left| \left(\frac{10^n}{2} + 1\right) \right. .$$

$a \neq \left(\frac{10^n}{2} + 1\right)$ as this would imply that $b = 2a = 10^n + 1$ is an $(n+1)$-digit number. So $a \neq 1$ is a proper divisor of $\left(\frac{10^n}{2} + 1\right)$. Since a is an n-digit number $n \geq 10^{n-1}$. So we have

$$\frac{1}{a}\left(\frac{10^n}{2} + 1\right) \leq \frac{1}{10^{n-1}}\left(\frac{10^n}{2} + 1\right) = 5 + \frac{1}{10^{n-1}}.$$

Since $n > 1$, $\frac{1}{a}\left(\frac{10^n}{2} + 1\right) \leq 5$ is an odd number. As 5 does not divide $\frac{1}{a}\left(\frac{10^n}{2} + 1\right)$ and $a \neq \left(\frac{10^n}{2} + 1\right)$, we must have $\frac{1}{a}\left(\frac{10^n}{2} + 1\right) = 3$. This gives

$a = \dfrac{10^n+2}{6}$, and so $b = 2a = \dfrac{10^n+2}{3}$. The following table gives the values of a and b for $n = 2, 3, 4, 5$:

n	10^n+2	$a = \frac{10^n+2}{6}$	$b = 2a$	$10^n a + b$
2	102	17	34	1734
3	1002	167	334	167334
4	10002	1667	3334	16673334
5	100002	16667	33334	1666733334

- **Case 3** $\dfrac{b}{a} = 4$, or $b = 4a$.
In this case, we have $a > 1$ for if $a = 1$ then $b = 4$ and $n = 1$. Then $b|(10^n a + b) \implies 4|14$. Now

$$ab|(10^n a + b) \implies 4a^2|(10^n a + 4a) \implies 4a|(10^n + 4) \implies a \,\Big|\, \left(\frac{10^n}{4} + 1\right).$$

Evidently $a \neq \left(\dfrac{10^n}{4} + 1\right)$ as otherwise $b = 4a$ will be a $(n+1)$-digit number. Since $n > 1$, $a \neq 1$. So

$$1 < \frac{1}{a}\left(\frac{10^n}{4} + 1\right) \leq \frac{1}{10^{n-1}}\left(\frac{10^n}{4} + 1\right) = \frac{10}{4} + \frac{1}{10^{n-1}}.$$

Since $a > 1 \implies n > 1$, the only possible value of $\dfrac{1}{a}\left(\dfrac{10^n}{4} + 1\right)$ is 2. Thus we have

$$\frac{1}{a}\left(\frac{10^n}{4} + 1\right) = 2, \quad \text{or } a = \frac{10^n}{8} + \frac{1}{2}.$$

Since a is an integer we must have $n = 2$. This gives $a = \dfrac{100}{8} + \dfrac{1}{2} = 13$ and $b = 4a = 52$.

- **Case 4** $\dfrac{b}{a} = 5$, or $b = 5a$.
In this case, a divides $\left(\dfrac{10^n}{5} + 1\right)$. Since a is an n-digit number we must have $a = \dfrac{10^n}{5} + 1$. But this would imply that $b = 5a = 10^n + 5$ which is a $(n+1)$-digit number. So there are no solutions in this case.

- **Case 5** $\dfrac{b}{a} = 8$, or $b = 8a$.

  In this case a divides $\left(\dfrac{10^n}{8} + 1\right)$. Again since a is an n-digit number we must have $a = \dfrac{10^n}{8} + 1$, but then b would be a $(n+1)$-digit number. Therefore, there are no solutions in this case as well.

  The following table gives a summary of what we have proved:

n	a	b
1	1	1
1	2	4
1	3	6
2	13	52
>1	$\frac{10^n+2}{6}$	$\frac{10^n+2}{3}$
$6k+3$	$\frac{10^n+1}{7}$	$\frac{10^n+1}{7}$

95. Let

$$\alpha = \frac{a + b\sqrt{-3}}{2}, \quad a, b \in \mathbb{Z}, \quad \text{both odd.}$$

Show that either $\omega\alpha \in \mathbb{Z}[\sqrt{-3}]$ or $\omega^2\alpha \in \mathbb{Z}[\sqrt{-3}]$, where $\omega = \frac{-1+\sqrt{-3}}{2}$ is a primitive cube root of unity.

Solution:

If a, b odd and $a \equiv b \pmod 4$ then

$$\left(\frac{a + b\sqrt{-3}}{2}\right)\omega = \frac{(-a - 3b) + \sqrt{-3}(a - b)}{4} \in \mathbb{Z}[\sqrt{-3}].$$

$a \equiv b \pmod 4 \Longrightarrow (-a - 3b) \equiv 0 \pmod 4$ and $(a - b) \equiv \pmod 4$.

If a, b odd and $a \not\equiv b \pmod 4$ then

$$\left(\frac{a + b\sqrt{-3}}{2}\right)\omega^2 = \frac{(-a + 3b) + \sqrt{-3}(a + b)}{4} \in \mathbb{Z}[\sqrt{-3}]$$

a, b odd and $a \not\equiv b \pmod 4 \Longrightarrow (-a + 3b) \equiv 0 \pmod 4$ and $(a + b) \equiv \pmod 4$.

96. Express the following (algebraic) integers in $\mathbb{Z}[\omega]$ (where $\omega = \dfrac{-1+\sqrt{-3}}{2}$ is a primitive cube root of unity) as a product of primes and using this prime factorization

(or otherwise) find their GCD. The prime factors of α and β that appear in your factorization should be elements of $\mathbb{Z}[\sqrt{-3}]$.

$$\alpha = \frac{175 + 77\sqrt{-3}}{2}, \qquad \beta = \frac{193 + 61\sqrt{-3}}{2}.$$

Your answer should be an element of $\mathbb{Z}[\sqrt{-3}]$ lying in the first or second quadrant.

Answer:

Note that $N(\alpha) = N(\beta) = 12103 = 7^3 \cdot 13 \cdot 19$.

The prime factorization of α is

$$-\omega(2 + \sqrt{-3})(1 + 2\sqrt{-3})(4 + \sqrt{-3})(2 - \sqrt{-3}).$$

The prime factorization of β is

$$-\omega^2 \cdot (2 - \sqrt{-3})(1 + 2\sqrt{-3})(4 - \sqrt{-3})(2 - \sqrt{-3}).$$

So the GCD of α and β is $8 + 3\sqrt{-3}$.

97. Find the GCD of the Gaussian integers $\alpha = 23 + 119i$, $\qquad \beta = 352 - 31i$ in $\mathbb{Z}[i]$ by expressing each of them as a product of prime numbers in $\mathbb{Z}[i]$. Your answer should be a Gaussian integer lying in the first quadrant.

Answer:

$$23 + 119i = (8 + 7i)(2 + i)(1 + i)(3 - 2i) = (-i)(8 + 7i)(2 + i)(1 + i)(2 + 3i)$$

$$352 - 31i = (-1)(8 + 7i)(1 + 2i)(4 + i)(2 + 3i)$$

$$\text{GCD} = (8 + 7i)(2 + 3i) = (-5 + 38i) \sim (38 + 5i).$$

98. Express the algebraic integers $\alpha = 54+105\sqrt{-2}$, $\qquad \beta = 106+52\sqrt{-2}$ in $\mathbb{Z}[\sqrt{-2}]$ as a product of primes and using this prime factorization find their GCD.

Your answer should be an element of $\mathbb{Z}[\sqrt{-2}]$ lying in the first or second quadrant.

Answer: The prime factorization of $\alpha = 54 + 105\sqrt{-2}$ is

$$(-1)(1 - \sqrt{-2})(1 + \sqrt{-2})(1 + 3\sqrt{-2})(1 + 6\sqrt{-2})\sqrt{-2}.$$

The prime factorization of $\beta = 106 + 52\sqrt{-2}$ is

$$\sqrt{-2}\sqrt{-2}(1 + \sqrt{-2})(1 + 3\sqrt{-2})(1 + 6\sqrt{-2}).$$

The GCD is

$$(52 - 53\sqrt{-2}) \sim (-52 + 53\sqrt{-2}).$$

99. Find the GCD of the following Gaussian integers. Your answer should be a Gaussian integer lying in the first quadrant. Provide the factorization into primes for each of the Gaussian integers appearing in the problems.

(a) $11 + 7i$ and $18 - i$.

(b) 85 and $1 + 13i$.

(c) $47 - 13i$ and $53 + 56i$.

Solution:

(a)

$$11 + 7i = (-i)(1+i)(1+2i)(4+i)$$

$$18 - i = (-i)(2+i)^2(3+2i)$$

So GCD $= 1$.

(b)

$$85 = (-i)(2+i)(2-i)(4+i)(4-i)$$

$$1 + 13i = (4+i)(1+i)(2+i)$$

So GCD $= (2+i)(4+i) = 7 + 6i$.

(c)

$$47 - 13i = (-1)(1+i)(2+5i)(4+5i)$$

$$53 + 56i = (-i)(1+2i)(5+2i)(4+5i)$$

So GCD $= 4 + 5i$.

100. Prove in detail that the only solutions of the Diophantine equation $y^2 + 11 = x^3$ are $x = 3,\ \ y = \pm 4$ and $x = 15,\ \ y = \pm 58$.

Solution:

If x is even then y will be odd. This is not possible as then $x^3 \equiv 0 \pmod 8$ and $y^2 + 11 \equiv 4 \pmod 8$. So for any possible solution y must be even and x odd. We know that $\mathbb{Q}(\sqrt{-11})$ is a Euclidean domain. So

$$D = \mathbb{Z}\left[\frac{-1+\sqrt{-11}}{2}\right] = \left\{\frac{a+b\sqrt{-11}}{2}, a \equiv b \pmod 2\right\}$$

is a UFD (unique factorization domain). The only units in D are ± 1. We have

$$(y+\sqrt{11}i)(y-\sqrt{11}i) = x^3.$$

We claim that $(y+\sqrt{11}i)$ and $(y-\sqrt{11}i)$ are relatively prime. Let $\alpha \in D$ be a divisor of each of these. Then

$$\alpha|(y+\sqrt{11}i), \ \ \alpha|(y-\sqrt{11}i) \Rightarrow N(\alpha)|(y^2+11) \Rightarrow N(\alpha) \ \text{ is odd.}$$

Also

$$\alpha|2\sqrt{11}i \Rightarrow N(\alpha)|44 \Rightarrow N(\alpha)|11 \Rightarrow N(\alpha) = 11 \ \text{ or } \ N(\alpha) = 1.$$

Since $N(\alpha)$ divides $4y^2$ it divides y^2. So $N(\alpha) = 11 \Rightarrow 11|y$, which is not possible since $y^2+11 = x^3$. Hence $N(\alpha) = 1$ and α is a unit. This shows that $(y+\sqrt{11}i)$ and $(y-\sqrt{11}i)$ are relatively prime. Since D is a UFD each of $(y+\sqrt{11}i)$ and $(y-\sqrt{11}i)$ is a cube in D. Let

$$y+\sqrt{11}i = \left(\frac{a+b\sqrt{11}i}{2}\right)^3, \quad a, b \in \mathbb{Z}, \quad a \equiv b \pmod 2.$$

Thus we have

$$8(y+\sqrt{11}i) = (a+b\sqrt{11}i)^3 = a(a^2-33b^2) + b(3a^2-11b^2)\sqrt{11}i$$

and so

$$8y = a(a^2-33b^2), \quad \text{and} \quad 8 = b(3a^2-11b^2).$$

Therefore b divides 8. We tabulate each of the possible values of b in the following table:

b	$3a^2-11b^2$	Equation for a	a	x	y
1	8	$3a^2 = 19$	No solution		
-1	-8	$3a^2 = 3$	± 1	3	∓ 4
2	4	$3a^2 48$	± 4	15	± 58
-2	-4	$3a^2 = 40$	No solution		
4	2	$3a^2 = 178$	No solution		
-4	2	$3a^2 = 174$	No solution		
8	1	$3a^2 705$	No solution		
-8	-1	$3a^2 703$	No solution		

Hence, the only solutions of the Diophantine equation $y^2+11=x^3$ are $x=3, y=\pm 4$ and $x=15, y=\pm 58$.

101. Prove that the Diophantine equation $y^2+5=x^3$ has no solutions.

Proof:

If x is even then y must be odd. In that case $x^3 \equiv 0 \pmod 8$ but $y^2+5 \equiv 6 \pmod 8$. So we assume that x is odd and y even. If $x \equiv 3 \pmod 4$ then $x^3 \equiv 3$, or $7 \pmod 8$. But $y^2+5 \equiv 5$ or $1 \pmod 8$. If $x \equiv 1 \pmod 4$ then $y^2+4=x^3-1=(x-1)(x^2+x+1)$. If $x \equiv 1 \pmod 4$ then $x^2+x+1 \equiv 3 \pmod 4$ and x^2+x+1 must have a prime divisor $p \equiv 3 \pmod 4$. For this prime $p \equiv 3 \pmod 4$, we have $1=\left(\frac{-4}{p}\right)=-1$.

102. Let $p \equiv 3 \pmod 3$ be a prime number. Prove that there are no rational numbers $x, y \in \mathbb{Q}$ such that $x^2+y^2=p$.

Solution:

It is enough to prove that the Diophantine equation $a^2+b^2=pc^2$ has no solutions. If $a^2+b^2=pc^2$, $p \equiv 3 \pmod 4$ has solutions in integers then

$$1=\left(\frac{-1}{p}\right)=-1.$$

Contradiction.

103. Let $a, b \in \mathbb{Z}$ be of different parity and $gcd(a, b)=1$. Prove that

$$gcd(a+bi, a-bi)=1 \text{ in } \mathbb{Z}[i]$$

Proof:

Let N denote the usual Euclidean norm function on $\mathbb{Z}[i]$ so $N(x+yi)=x^2+y^2$, for $x, y \in \mathbb{Z}[i]$.

Since a and b have different parity a^2+b^2 is odd. Let $\alpha \in \mathbb{Z}[i]$ be such that $\alpha|(a+bi)$ and $\alpha|(a-bi)$. As $\alpha|(a+bi)$ implies that $N(\alpha)|(a^2+b^2)$ therefore $N(\alpha)$ is odd. Also

$$\alpha|(a \pm bi) \Longrightarrow \alpha|2a, \alpha|2bi \Longrightarrow N(\alpha)|4a^2, \text{ and } N(\alpha)|4b^2.$$

As $N(\alpha)$ is odd and $gcd(a, b)=1$, we conclude that $N(\alpha)=1$ so $gcd(a+bi, a-bi)=1$.

104. Show that the only solutions of the Diophantine equation $y^2 + 4 = x^3$ are $(x, y) = (5, \pm 11))$ and $(x, y) = (2, \pm 2)$

Solution:

- **Case I** y is odd.

  In this case, we have $(y+2i)(y-2i) = x^3$. By previous problem $gcd(y+2i, y-2i) = 1$. As $\mathbb{Z}[i]$ is a unique factorization domain, we have $(y+2i) = (a+bi)^3$ for some $a, b \in \mathbb{Z}$. We have $y+2i = a^3-3abi(a+bi)+(ib)^3 = a^3-3ab^2+i(3a^2b-b^3)$. This gives $y = a^3 - 3ab^2$ and $3a^2b - b^3 = b(3a^2 - b^2) = 2$. Therefore $b = \pm 1$ or $b = \pm 2$. Now

  $$b = \pm 1 \Longrightarrow 3a^2 - 1 = \pm 2 \Longrightarrow 3a^2 = 1 \pm 2$$

  which is not possible. Again

  $$b = \pm 2 \Longrightarrow 3a^2 - 4 = \pm 1 \Longrightarrow 3a^2 = 3 \Longrightarrow a = \pm 1.$$

  This gives $y = a(a^2 - 3b^2) = \mp(1 - 12) = \pm 11$ and so $x = 5$.

- **Case 2** y is even.

  Let $y = 2z$. Then we have $4z^2+4 = x^3$. Let $x = 2t$. Then we have $4z^2+4 = 8t^3$ or $z^2 + 1 = 2t^3$. So z is odd. We have

  $$\left(\frac{z+1}{2}\right)^2 + \left(\frac{z-1}{2}\right)^2 = \frac{z^2+1}{2} = t^3.$$

  Let $a = \dfrac{z+1}{2}$ and $b = \dfrac{z-1}{2}$. Then we have $a^2+b^2 = t^3$ or $(a+bi)(a-bi) = t^3$. Since $a - b = 1$, $gcd(a, b) = 1$ and a, b have different parity. Again by the previous problem $gcd(a + bi, a - bi) = 1$. By the unique factorization property in $\mathbb{Z}[i]$ we have $a + bi = (u + iv)^3$, $u, v \in \mathbb{Z}$. Now

$$a + bi = (u + iv)^3 = u^3 + 3iuv(u + iv) + (iv)^3 = u^3 - 3uv^2 + i(3u^2v - v^3).$$

  This gives $a = u^3 - 3uv^2$ and $b = 3u^2v - v^3$. Since $a - b = \dfrac{z+1}{2} - \dfrac{z-1}{2} = 1$, we have

  $$\begin{aligned} u^3 - 3uv^2 - (3u^2v - v^3) &= 1 \\ u^3 + v^3 - 3uv(u + v) &= 1 \end{aligned}$$

$$(u+v)(u^2-uv+v^2)-3uv(u+v)=1$$
$$(u+v)(u^2-4uv+v^2)=1.$$

So $u+v=\pm 1$ and $u^2-4uv+v^2=\mp 1$. Now

$$u^2-4uv+v^2=(u+v)^2-6uv=\mp 1 \Longrightarrow 1-6uv=\mp 1 \Longrightarrow uv=0$$

$$u=0 \Longrightarrow a=0 \Longrightarrow z=-1 \Longrightarrow y=-2, x=2$$

and

$$v=0 \Longrightarrow b=0 \Longrightarrow z=1 \Longrightarrow y=2, x=2.$$

So the only solutions in case are $(x,y)=(2,\pm 2)$.

105. Using that $\mathbb{Z}[\sqrt{-2}]$ is a unique factorization domain prove that the only solutions of the Diophantine equation $y^2+2=x^3$ are $(x,y)=(3,\pm 5)$.

Let N denote the Euclidean norm function on $\mathbb{Z}[\sqrt{-2}]$, so $N(a+b\sqrt{-2})=a^2+2b^2$. If y is even then x is even and $2=x^3-y^2\equiv 0 \pmod 4$ which is not true. So y is odd. We have $(y+\sqrt{-2})(y-\sqrt{-2})=x^3$. We claim that $gcd(x+\sqrt{-2},x-\sqrt{-2})=1$. Let $\alpha\in\mathbb{Z}[\sqrt{-2}]$ be such that α divides both $y+\sqrt{-2}$ and $y-\sqrt{-2}$. So $N(\alpha)$ divides y^2+2 and so $N(\alpha)$ is odd. Also

$$\alpha|(y+\sqrt{-2}) \text{ and } \alpha|(y-\sqrt{-2}) \Longrightarrow \alpha|2\sqrt{-2} \Longrightarrow N(alpha)|8.$$

Thus $N(\alpha)=1$ and $gcd(x+\sqrt{-2},x-\sqrt{-2})=1$. Now we have $y+\sqrt{-2}=(a+b\sqrt{-2})^3$ for some $a,b\in\mathbb{Z}$. This gives

$$y+\sqrt{-2}=a^3+3ab\sqrt{-2}(a+b\sqrt{-2})+(b\sqrt{-2})^3=a^3-6ab^2+(3a^2b-2b^3)\sqrt{-2}$$

and we have $y=a^3-6ab^2$ and $1=3a^2b-2b^3$ or $b(3a^2-2b^2)=1$. Therefore $b=\pm 1$. Now

$$b=-1 \Longrightarrow 3a^2-2=-1 \Longrightarrow 3a^2=1$$

which is not possible. So $b=1$ which gives $a=\pm 1$. Therefore $y=\pm 5$ and $x=3$.

106. Let $f(x)=x^4+1$. Prove the following:

(a) $f(x)$ factors as product of four linear factors mod p if $p=2$ or $p\equiv 1 \pmod 8$.

(b) $f(x)$ factors as product of two quadratic factors mod p if $p\equiv 3,5$ or $7 \pmod 8$.

Solution:

(a) If $p = 2$ then $x^4 + 1 \equiv (x+1)^4 \pmod 2$.

Let $p \equiv 1 \pmod 8$. Then $\left(\frac{2}{p}\right) = \left(\frac{-2}{p}\right) = 1$. Let $a \in \mathbb{Z}$ be such that $a^2 \equiv 2 \pmod p$. We have

$$x^4+1 = x^4+2x^2+1-2x^2 = (x^2+\sqrt{2}x+1)(x^2-\sqrt{2}x+1) \equiv (x^2+ax+1)(x^2-ax+1) \pmod p.$$

Now $\text{Discriminant}(x^2+\sqrt{2}x+1) = \text{Discriminant}(x^2-\sqrt{2}x+1) = 2-4 = -2$ Since $\left(\frac{-2}{p}\right) = 1$ both $(x^2+\sqrt{2}x+1)$ and $(x^2+\sqrt{2}x+1)$ can be factored into linear factors mod p.

(b)

- Let $p \equiv 7 \pmod 8$. Since $\left(\frac{2}{p}\right) = 1$, $f(x) = x^4+1 \equiv (x^2+ax+1)(x^2-ax+1) \pmod p$ where $a^2 \equiv 2 \pmod p$. Since $\left(\frac{-2}{p}\right) = -1$ both of these quadratic factors are irreducible mod p.
- Let $p \equiv 3 \pmod 8$. Then we have

$$x^4+1 = x^4-2x^2+1+2x^2 = (x^2-1)^2-(\sqrt{-2}x)^2 = (x^2+\sqrt{-2}x-1)(x^2-\sqrt{-2}x-1) \pmod p.$$

Since $\left(\frac{-2}{p}\right) = 1$, $\left(\frac{2}{p}\right) = -1$, and

$$\text{Discriminant}(x^2+\sqrt{-2}x-1) = \text{Discriminant}(x^2-\sqrt{-2}x-1) = -2+4 = 2$$

x^4+1 factors as a product of two irreducible quadratics mod p.

- Let $p \equiv 5 \pmod 8$. We show that in this case we can find $a, b \in \mathbb{Z}$ such that $x^4+1 \equiv (x^2+a)(x^2+b) \pmod p$ where both x^2+a and x^2+b are irreducible mod p. Since $\left(\frac{-1}{p}\right) = 1$ we can find $a \in \mathbb{Z}$ such that $a^2 \equiv -1 \pmod p$. Let $b \in \mathbb{Z}$ be such that $ab \equiv 1 \pmod p$ or $b \equiv a^{-1} \pmod p$. Then $a+b = a+a^{-1} \equiv \frac{a^2+1}{a} \equiv b(a^2+1) \equiv 0 \pmod p$. Then we have $(x^2+a)(x^2+b) = x^4+(a+b)+ab \equiv x^4+1 \pmod p$. We claim that x^2+a and x^2+b are irreducible mod p. For if x^2+a were reducible mod p then there would be some $c \in \mathbb{Z}$ such that $c^2 \equiv a \pmod p$ or $c^4 \equiv a^2 \equiv -1 \pmod p$. Since $\frac{p-1}{4}$ is odd this would imply that

$$1 \equiv c^{p-1} = \left(c^4\right)^{\frac{p-1}{4}} \equiv \left(a^2\right)^{\frac{p-1}{4}} \equiv (-1)^{\frac{p-1}{4}} \equiv -1 \pmod p$$

which is not true.

Factoring $x^4 + 1$ mod p.

Prime p	Factors of $x^4 + 1$ modulo p	p (mod 8)
2	$(x+1)^4$	
17	$(\boldsymbol{x}+\mathbf{2})(\boldsymbol{x}+\mathbf{8})(\boldsymbol{x}+\mathbf{9})(\boldsymbol{x}+\mathbf{15})$	1
41	$(\boldsymbol{x}+\mathbf{3})(\boldsymbol{x}+\mathbf{14})(\boldsymbol{x}+\mathbf{27})(\boldsymbol{x}+\mathbf{38})$	1
3	$(x^2+x+2)(x^2+2x+2)$	3
11	$(x^2+3x+10)(x^2+8x+10)$	3
5	$(x^2+2)(x^2+3)$	5
13	$(x^2+5)(x^2+8)$	5
7	$(x^2+3x+1)(x^2+4x+1)$	7
23	$(x^2+5x+1)(x^2+18x+1)$	7

107. Suppose that

$$\frac{EVE}{DID} = .TALKTALKTALKTALKTALK.....$$

where the left-hand side represents a common fraction (in reduced form) and each of the letter represents a different digit. Find these digits.

Solution:

We have

$$\begin{aligned} \frac{EVE}{DID} &= .TALKTALKTALKTALKTALK..... \\ &= \frac{TALK}{9999} \\ &= \frac{TALK}{3 \cdot 3 \cdot 11 \cdot 101}. \end{aligned}$$

So we now have

$$\frac{EVE}{DID} = \frac{TALK}{3 \cdot 3 \cdot 11 \cdot 101}.$$

Since LHS is given to be a reduced fraction, the number represented by DID is a three digit number dividing $3 \cdot 3 \cdot 11 \cdot 101$. This gives only three possibilities for DID, namely 101, 909, and 303. We show that the first two are not possible.

Suppose that $DID = 101$. So $D = 1$ and $I = 0$. We have

$$\frac{EVE}{101} = \frac{TALK}{9999}.$$

Since $TALK < 9999$ we have $EVE < 101$. This forces the only choice for EVE to be 100 which is not possible (as E and V are different digits).

Next suppose that $DID = 909$. This gives $EVE \cdot 11 = TALK$. But this forces $E = K$, which we are assuming to not be the case.

Finally let $DID = 303$. This gives $EVE \cdot 33 = TALK$. Since $TALK < 9999$ we have

$$EVE = \frac{TALK}{33} < \frac{9999}{33} = 303.$$

So EVE must be of the form $1\Box1$ or $2\Box2$, where $\Box$ is a digit which has not been used thus far. As before $EVE \times 1\Box1 = TALK$ would imply that $E = K$ which we are assuming to not to be the case. So we are now left with $2\Box2$ as the only possible value for EVE. This means that the only possible values for EVE are $212, 242, 252, 262, 272, 282$, and 292. In view of the fact that $EVE \times 33 = TALK$, all excepting 242 are inadmissible as seen from the following table:

$$212 \times 33 = 6996$$
$$\mathbf{242} \times \mathbf{33} = \mathbf{7986}$$
$$252 \times 33 = 8316$$
$$262 \times 33 = 8646$$
$$272 \times 33 = 8976$$
$$282 \times 33 = 9306$$
$$292 \times 33 = 9636$$

This gives the only solution as $EVE = 242$, $DID = 303$ and $TALK = 7986$. We can check that

$$\frac{242}{303} = .7986\mathbf{7986}7986...$$

108. Prove that

$$\tan\frac{3\pi}{11}+4\sin\frac{2\pi}{11}=\sqrt{11}$$

Proof (Bromwich's Infinite Series, page 270):

Let $\alpha=e^{\frac{2\pi i}{11}}=\cos\frac{2\pi}{11}+i\sin\frac{2\pi}{11}$. Then $\alpha^{11}=1$ and

$$\alpha+\alpha^2+\alpha^3+\alpha^4+\alpha^5+\alpha^6+\alpha^7+\alpha^8+\alpha^9+\alpha^{10}=-1.$$

Let

$$A=\alpha+\alpha^4+\alpha^5+\alpha^9+\alpha^3,\qquad B=\alpha^2+\alpha^8+\alpha^{10}+\alpha^7+\alpha^6.$$

Then $A+B=-1$ and $A\cdot B=3$, and so A and B are the roots of the equation $x^2+x+3=0$. Since $\sin\frac{4\pi}{11}<\sin\frac{5\pi}{11}$ the imaginary part of $\alpha^2+\alpha^8$ is negative. Also same is true for the imaginary parts of α^{10}, α^7, and α^6. Therefore

$$A=\frac{-1+i\sqrt{11}}{2},\qquad B=\frac{-1-i\sqrt{11}}{2},\qquad A-B=i\sqrt{11}.$$

Now

$$i\tan\frac{3\pi}{11}=-i\tan\frac{8\pi}{11}=-\frac{2i\sin\frac{8\pi}{11}}{2\cos\frac{8\pi}{11}}=-\frac{\alpha^4-\alpha^{-4}}{\alpha^4+\alpha^{-4}}=\frac{\alpha^{-8}-1}{\alpha^{-8}+1}=\frac{\alpha^3-1}{\alpha^3+1}=\frac{\alpha^3-\alpha^{33}}{\alpha^3+1}=$$

$$-\frac{\alpha^3(\alpha^{30}-1)}{\alpha^3+1}=-\frac{\alpha^3(\alpha^{15}-1)(\alpha^{15}+1)}{\alpha^3+1}=-\alpha^3(\alpha^4-1)(\alpha^{12}-\alpha^9+\alpha^6-\alpha^3+1)=$$

$$(\alpha+\alpha^3+\alpha^4+\alpha^5+\alpha^9)-(\alpha^2+\alpha^6+\alpha^7+\alpha^8+\alpha^{10})-2(\alpha-\alpha^{10})=A-B-4i\sin\frac{2\pi}{11}=i\sqrt{11}-4i\sin\frac{2\pi}{11}.$$

So

$$\tan\frac{3\pi}{11}=\sqrt{11}-4\sin\frac{2\pi}{11}\quad\text{or}\quad\tan\frac{\mathbf{3\pi}}{\mathbf{11}}+\mathbf{4}\sin\frac{\mathbf{2\pi}}{\mathbf{11}}=\sqrt{\mathbf{11}}.$$

109. Find the quadratic irrational α whose continued fraction expansion is $\left(6;\overline{2,7}\right)$.

Solution:

We have

$$\alpha=\left(6;\overline{2,7}\right)=6+\frac{1}{2+}\;\frac{1}{7+}\cdots$$

So

$$\alpha-6=\frac{1}{2+\dfrac{1}{7+(\alpha-6)}}=\frac{1}{2+\dfrac{1}{\alpha+1}}=\frac{\alpha+1}{2\alpha+3}.$$

Therefore $(\alpha - 6)(2\alpha + 3) = \alpha + 1$ which gives $2\alpha^2 - 10\alpha - 19 = 0$. Since $\alpha > 0$, we have

$$\alpha = \frac{10 + \sqrt{100 + 152}}{4} = \frac{10 + \sqrt{252}}{4} = \frac{5 + 3\sqrt{7}}{2}.$$

110. Find the continued fraction expansion of the quadratic irrational $\alpha = \dfrac{3 + 5\sqrt{7}}{2}$.

Answer: $\left(8; \overline{8, 1, 2, 1, 8, 13}\right)$

111. Find the quadratic irrational α whose continued fraction expansion is $\left(3; \overline{1, 4, 1, 1}\right)$.

Answer: $\alpha = \dfrac{5 + \sqrt{7}}{2}$

112. Prove that the product of k consecutive natural numbers is divisible by $k!$.

Proof:

It is an immediate consequence of the following:

$$\binom{n+k}{k} = \frac{(n+1)(n+2)\cdots(n+k)}{k!}.$$

113. Prove that the Diophantine equation $13x^2 + 10xy + 2y^2 = 19380$ has no solutions.

Proof:

We note that $19380 \equiv 0 \pmod 3$ but $19380 \not\equiv 0 \pmod 9$. If $x, y \in \mathbb{Z}$ satisfy $13x^2 + 10xy + 2y^2 = 19380$ then $x \equiv 0 \pmod 3 \iff y \equiv 0 \pmod 3$. But then $13x^2 + 10xy + 2y^2 \equiv 0 \pmod 9$. So we assume that $x, y \not\equiv 0 \pmod 3$. Therefore $x^2 \equiv y^2 \equiv 1 \pmod 3$. Thus we have

$$19380 = 13x^2 + 10xy + 2y^2 \equiv 1 + xy + 2 \equiv xy \pmod 3 \Longrightarrow xy \equiv 0 \pmod 3$$

which is not true.

114. Prove that the Diophantine equation $y^2 = x^3 - 3$ has no solutions.

Proof:

Suppose that the given equations has solutions. Then x must be odd as otherwise $y^2 = x^3 - 3 \equiv 5 \pmod 8$ which is not possible. Also if $x \equiv 1 \pmod 4$ then $y^2 = x^3 - 3 \equiv 2 \pmod 4$ which too is not possible. Now suppose that $x \equiv 3 \pmod 4$ then we write $y^2 + 4 = x^3 + 1 = (x + 1)(x^2 - x + 1)$ and $x^2 - x + 1 \equiv 3 \pmod 4$. Let $p \equiv 3 \pmod 4$ be a primes divisor of $x^2 - x + 1$. Then as $y^2 + 4 \equiv 0 \pmod p$ we have

$$1 = \left(\frac{-4}{p}\right) = \left(\frac{-1}{p}\right) = -1.$$

Contradiction.

115. Using that the ring of integers in the imaginary quadratic field $\mathbb{Q}(\sqrt{-19})$ is a unique factorization domain show that the only solutions of the Diophantine equation $y^2 = x^3 - 19$ are $(x, y) = (7, \pm 18)$.

Solutions:

First, we recall that as discussed in the text the ring of integers in $\mathbb{Q}(\sqrt{-19})$ is

$$D = \left\{ \frac{a + b\sqrt{-19}}{2} : a, b \in \mathbb{Z}, a \equiv b \pmod{2} \right\}.$$

Suppose that the given equation has solutions. Then x cannot be even as x even implies that $y^2 = x^3 - 19 \equiv 5 \pmod{8}$, which is not possible. So x must be odd and $x \equiv 3 \pmod{4}$ because $x \equiv 1 \pmod{4}$ implies that $y^2 = x^3 - 19 \equiv 2 \pmod{4}$ which is not possible. We write $y^2 + 19 = x^3$ or $(y + \sqrt{-19})(y - \sqrt{-19}) = x^3$. We claim that $GCD(y + \sqrt{-19}, y - \sqrt{-19}) = 1$. Suppose $\alpha \in D$ divides both $y + \sqrt{-19}$ and $y - \sqrt{-19}$. This implies that α divides $2y$ and $2\sqrt{19}$. Therefore $N(\alpha)$ divides 76. As x is odd so is $N(\alpha)$ and therefore $N(\alpha) = 19$ or $N(\alpha) = 1$. But $N(\alpha) = 19$ implies that 19 divides x and therefore it also divides y which is not possible. So $N(\alpha) = 1$ and $\alpha = \pm 1$ is a unit in D.

Since D is a unique factorization domain there exists $\eta \in D$ such that $y + \sqrt{-19} = \eta^3$, where $\eta = \dfrac{a + b\sqrt{-3}}{2}$ and $a, b \in \mathbb{Z}$ have same parity. We have

$$y + \sqrt{-19} = \left(\frac{a + b\sqrt{-19}}{2} \right)^3 = \frac{a^3 - 57ab^2 + (3a^2b - 19b^3)\sqrt{-19}}{8}$$

or

$$8(y + \sqrt{-19}) = a^3 - 57ab^2 + (3a^2b - 19b^3)\sqrt{-19}.$$

This gives $8y = a^3 - 57ab^2$ and $8 = b(3a^2 - 19b^2)$. If both a and b are even then $b = \pm 2$ which implies that $3a^2 - 76 = \pm 4$ which is not possible. Therefore both a and b must be odd and $b = \pm 1$. Now $b = -1$ implies that $3a^2 = 11$ which is not possible. So the only possibility is $b = 1$ and therefore $3a^2 = 27$. This gives $a = \pm 3$. As $8y = a^3 - 57ab^2$, $b = 1, a = 3$ implies that $y = -18$ and $x = 7$. Also $b = 1, a = -3$ implies that $y = 18$ and $x = 7$.

116. Prove that the Diophantine equation $y^2 = x^3 - 3$ has no solutions.

Proof:

Suppose that the given equations has solutions. Then x must be odd as otherwise $y^2 = x^3 - 3 \equiv 5 \pmod 8$ which is not possible. Also if $x \equiv 1 \pmod 4$ then $y^2 = x^3 - 3 \equiv 2 \pmod 4$ which too is not possible. Now suppose that $x \equiv 3 \pmod 4$ then we write $y^2 + 4 = x^3 + 1 = (x+1)(x^2 - x + 1)$ and $x^2 - x + 1 \equiv 3 \pmod 4$. Let $p \equiv 3 \pmod 4$ be a primes divisor of $x^2 - x + 1$. Then as $y^2 + 4 \equiv 0 \pmod p$ we have

$$1 = \left(\frac{-4}{p}\right) = \left(\frac{-1}{p}\right) = -1.$$

Contradiction.

117. Using that the ring of integers in the imaginary quadratic field $\mathbb{Q}(\sqrt{-19})$ is a unique factorization domain show that the only solutions of the Diophantine equation $y^2 = x^3 - 19$ are $(x, y) = (7, \pm 18)$.

Solutions:

First, we recall that as discussed in the text the ring of integers in $\mathbb{Q}(\sqrt{-19})$ is

$$D = \left\{\frac{a + b\sqrt{-19}}{2} : a, b \in \mathbb{Z}, a \equiv b \pmod 2\right\}.$$

Suppose that the given equation has solutions. Then x cannot be even as x even implies that $y^2 = x^3 - 19 \equiv 5 \pmod 8$, which is not possible. So x must be odd and $x \equiv 3 \pmod 4$ because $x \equiv 1 \pmod 4$ implies that $y^2 = x^3 - 19 \equiv 2 \pmod 4$ which is not possible. We write $y^2 + 19 = x^3$ or $(y + \sqrt{-19})(y - \sqrt{-19}) = x^3$. We claim that $GCD(y + \sqrt{-19}, y - \sqrt{-19}) = 1$. Suppose $\alpha \in D$ divides both $y + \sqrt{-19}$ and $y - \sqrt{-19}$. This implies that α divides $2y$ and $2\sqrt{19}$. Therefore $N(\alpha)$ divides 76. As x is odd so is $N(\alpha)$ and therefore $N(\alpha) = 19$ or $N(\alpha) = 1$. But $N(\alpha) = 19$ implies that 19 divides x and therefore it also divides y which is not possible. So $N(\alpha) = 1$ and $\alpha = \pm 1$ is a unit in D.

Since D is a unique factorization domain, there exists $\eta \in D$ such that $y + \sqrt{-19} = \eta^3$, where $\eta = \dfrac{a + b\sqrt{-3}}{2}$ and $a, b \in \mathbb{Z}$ have same parity. We have

$$y + \sqrt{-19} = \left(\frac{a + b\sqrt{-19}}{2}\right)^3 = \frac{a^3 - 57ab^2 + (3a^2b - 19b^3)\sqrt{-19}}{8}$$

or

$$8(y+\sqrt{-19}) = a^3 - 57ab^2 + (3a^2b - 19b^3)\sqrt{-19}.$$

This gives $8y = a^3 - 57ab^2$ and $8 = b(3a^2 - 19b^2)$. If both a and b are even then $b = \pm 2$ which implies that $3a^2 - 76 = \pm 4$ which is not possible. Therefore, both a and b must be odd and $b = \pm 1$. Now $b = -1$ implies that $3a^2 = 11$ which is not possible. So the only possibility is $b = 1$ and therefore $3a^2 = 27$. This gives $a = \pm 3$. As $8y = a^3 - 57ab^2$, $b = 1, a = 3$ implies that $y = -18$ and $x = 7$. Also $b = 1, a = -3$ implies that $y = 18$ and $x = 7$.

118. Prove that the Diophantine equation $x^4 + y^3 = 13z + 7$ has no solutions.

Proof:

We consider the given equation mod 13. Since 2 is a primitive root mod 13 the possible fourth power mod 13 are $0, 2^4, 2^8, 2^{12}$. So the possible values of $x^4 \pmod{13}$, are $0, 1, 3, 9$. Likewise the possible third powers mod 13 are $0, 2^3, 2^6, 2^9, 2^{12}$ and so the possible values of $y^3 \pmod{13}$ are $0, 1, 5, 8, 12$. None of the possible values of $x^4 \pmod{13}$ and $y^3 \pmod{13}$ add up to $7 \pmod{13}$. Therefore, the given Diophantine equation has no solutions.

119. Prove that the following Diophantine equations have no solutions:

(a) $x^4 + y^4 = 17z + 11$.

(b) $x^5 + y^2 = 11z + 7$.

(c) $x^6 + y^3 = 19z + 5$.

Proof:

The proof of these is very similar to the one in the previous problem.

120. Consider all of natural numbers written as a long string of digits such as $123456789101112131415161718192021\cdots$. That is write **one, two, three, four, five, six, seven, eight, nine, ten, eleven** $\cdots$ and replace each of these with their decimal representation forming a long unending string of digits (without commas). The *Beginning Position* of a natural number N in this string is defined to be the position that the first digit of N occupies in the string. For example, the Beginning Position for 21 is 32.

- What is the Beginning Position for the number $N = 2020$.
- Express the Beginning Position of a natural number N as a function of N.

Answer:

- 6970.
- The **Beginning Position** for N is $\sum_{i=1}^{k} i(10^i - 10^{i-1}) + (k+1)(N - 10^k) + 1$, where $k = [\log_{10} N]$ or,

$$\text{Beginning Position(N)} = \sum_{i=1}^{[\log_{10} N]} i(10^i - 10^{i-1}) + ([\log_{10} N] + 1)\left(N - 10^{[\log_{10} N]}\right) + 1.$$

121. Show that if $p > 2$ is a prime number then the congruence $x^4 + 4 \equiv 0 \pmod{p}$ has solutions if and only if $p \equiv 1 \pmod{4}$.

Solution:

We have

$$x^4 + 4 = x^4 + 4x2 + 4 - 4x^2 = (x^2+2)^2 - (2x)^2 = (x^2 + 2x + 2)(x^-2x + 2)$$

and Discriminant$(x^2 + 2x + 2)$ = Discriminant$(x^2 - 2x + 2) = -4$. So $x^4 + 4 \equiv 0 \pmod{p}$ has solutions if and only if $\left(\dfrac{-4}{p}\right) = \left(\dfrac{-1}{p}\right) = 1$.

122. Solve the congruence $(x^2 + 1)(x + 1)x \equiv -1 \pmod{61}$.

Answer:

Note that 2 is a primitive root modulo 61. So the solutions are $2^{12}, 2^{24}, 2^{36}, 2^{48} \pmod{61}$. That is $9, 20, 58, 34 \pmod{61}$.

123. Decrypt the message P L G) R F (F I ! which has been encrypted using the RSA encryption method as discussed in the text. The particulars of the encryption are as follows:

(a) $n = 36919997$.

(b) $e = 26890259$.

(c) The sender's block size $k = 4$. The receiver's block size $\ell = 5$.

(d) The 34 characters are used in the formation of the numerical equivalent. The 26 Roman letters "A" to "Z" are numbered from zero to 25. Blank space= 26, period = 27, comma = 28, and the exclamation (!) = 29, semicolon = 30, hyphen = 31, left-paren = 32, and right-paren = 33.

A	B	C	D	E	F	G	H	I	J	K	L	M	N	O	P	Q
0	1	2	3	4	5	6	7	8	9	10	11	12	13	14	15	16

R	S	T	U	V	W	X	Y	Z	□	.	,	!	;	–	(	)
17	18	19	20	21	22	23	24	25	26	27	28	29	30	31	32	33

Solution:

$$n = 36919997 = p \cdot q = 5839 \cdot 6323, \qquad \phi(n) = 36907836 \qquad d = 5987$$

$$\text{PLG)R} = (15, 11, 6, 33, 17) \equiv 20485459 \qquad \text{F(FI!} = (5, 32, 5, 8, 29) \equiv 7945489$$

$$20485459^{5987} \pmod{36919997} = 763484 = (19, 14, 15, 14) = \text{T O P O}$$

$$7945489^{5987} \pmod{36919997} = 448756 = (11, 14, 6, 24) = \text{L O G Y.}$$

So the message is T O P O L O G Y.

Encryption:

TOPO $= (19, 14, 15, 14) \equiv 763484.$

LOGY $= (11, 14, 6, 24) \equiv 448756.$

$$763484^{26890259} \pmod{36919997} = 20485459 = (15, 11, 6, 33, 17) = \text{P L G) R}$$

$$448756^{26890259} \pmod{36919997} = 7945489 = (5, 32, 5, 8, 29) = \text{F (F I !}$$

124. Decrypt the following message:

IQJFA□CPVIBWBALBPV

The particulars of the encryption are as follows:

(a) $n = 7081$.

(b) $e = 233$.

(c) The sender's block size $k = 2$. The receiver's block size $\ell = 3$.

(d) The 27 characters are used in the formation of the numerical equivalent. The 26 roman letters "A" to "Z" are numbered from 0 to 25. Blank space = 26 as indicated in the following table:

A	B	C	D	E	F	G	H	I	J	K	L	M	N	O	P	Q	R	S	T	U	V	W	X	Y	Z	□
0	1	2	3	4	5	6	7	8	9	10	11	12	13	14	15	16	17	18	19	20	21	22	23	24	25	26

Solution

Input Block	Integer Digits	x	$y = x^d \pmod n$	Integer Digits of y	Output Block
IQJ	$(8, 6, 9)$	6273	53	$(1, 26)$	$B\square$
$FA\square$	$(5, 0, 26)$	3671	395	$(14, 17)$	OR
CPV	$(2, 15, 21)$	1884	702	$(26, 0)$	$\square A$
IBW	$(8, 1, 22)$	5881	714	$(26, 12)$	$\square M$
BAL	$(1, 0, 11)$	740	229	$(8, 13)$	IN
BPV	$(1, 15, 21)$	1155	558	$(20, 18)$	US

Decrypted Message: B OR A MINUS.

125. Assume that the plain text consists of two-letter blocks in the usual 26-letter alphabet (converted to an integer between 0 and 26^2—1 in the usual way) and the cipher text consists of three-letter block in the same alphabet. Decrypt the following message

AVEAXSBBP

assuming that the message was encrypted using the decryption key $(n, e) = (2537, 1271)$:

Answer: $2537 = 43 \times 59$, $d = 23$. Decrypted Message is **FERMAT**.

126. Assume that the plain text consists of two-letter blocks in the usual 26-letter alphabet (converted to an integer between 0 and 26^2—1 in the usual way) and the cipher text consists of three-letter block in the same alphabet. Perform the following *Encryption* and *Decryption*:

(a) Encrypt the message **LOVE** using the encryption key $(n, e) = (1147, 43)$.

(b) Decrypt the following message **BHFBMD** using the decryption key $(n, d) = (1189, 19)$.

Answer:

(a) **ANZ BDF**

(b) **PASS**

127. Suppose that p is a prime which can be represented by any one of the forms x^2+14y^2, $2x^2+7y^2$, $3x^2 \pm 2xy + 5y^2$ then $\left(\frac{-14}{p}\right) = 1$.

Solution:

$$x^2+14y^2 = p \Longrightarrow \left(\frac{-14}{p}\right) = 1.$$

$$2x^2+7y^2 = p \Longrightarrow 4x^2+14y^2 = 2p \Longrightarrow \left(\frac{-14}{p}\right) = 1.$$

$$3x^2 \pm 2xy + 5y^2 = p \Longrightarrow 9x^2 \pm 6xy + 15y^2 = 3p \Longrightarrow (3x \pm y)^2 + 14y^2 = 3p \Longrightarrow \left(\frac{-14}{p}\right) = 1.$$

Credits and Sources Acknowledged

> *As for credits—who discovered what?, who was the first?, whose solution is the best? ⋯ It wouldn't be fair to give credit in some cases and not in others. We are all pretty sure that Pythagoras did not discover the Pythagorean theorem, and it doesn't really matter, does it? The beauty of the mathematics speaks for itself, and so be it. Halmos (1991)*

As mentioned in the introduction during the course of our long teaching career, we had the opportunity to look at and benefit from the available books on number theory and related mathematics literature. Some of the problems whose solutions are included in this book have appeared in somewhat modified form in one or more books listed in the bibliography. Some have been suggested by colleagues, some are folklore, and some we had created based on our interaction with the students. Below is only a partial list of the sources that we specifically remember and recognize:

- For the chapter on Continued Fractions (Davenport, 1993)
- For Chapter 6 (Nagell, 1981)
- For discussion of RSA Cryptography in Chapter 10 (Koblitz, 1987)
- Problem # 93 (Niven et al., 1991)
- Problems # 60, 61 ICM 2014 Seoul Newsletters
- Problems # 10, 120 Museum of Mathematics, New York
- Problems # 69, 94 Personal communication by a friend/colleague
- Problem # 104 (Uspensky and Heaslet, 1939)
- Problems # 23, 39, 86, 107 (Beiler, 1966)
- Problem # 108 (Bromwich, 1991; Williams and Hardy, 1996)
- Problem # 45 (Hardy and Wright, 1960)
- Problem # 26 (Burn and Chetwynd, 1996)
- Concerning the **Prime Charts** in appendix, we learned about the Prime Number Cards for primes up to 2500 (except for the primes 2 and 5) from *mathcircles.org*

and we were told to contact Dr. Paul Zeitz at the University of San Francisco who in turn told us that a student of his gave this card to him many years ago but he has been disseminating it for years. For the extension of these Prime Number Cards to primes up to 10,000, we thank our student Joel Kamm.

- For converting our PowerPoint slides into eps files, we thank our student Christopher Hegland.

Glossary

- **Algebraically Closed Field**

  A field $\mathbb{K}$ is said to be algebraically closed if for every polynomial $f(x) = a_0x^n + a_1x^{n-1} + \cdots\cdots a_{n-1}x + a_n \in \mathbb{K}[x]$, $n > 0$ there exist $\alpha_i \in \mathbb{K}$, $1 \leq i \leq n$ such that

$$f(x) = a_0x^n + a_1x^{n-1} + \cdots\cdots a_{n-1}x + a_n = a_0\prod_{i=1}^{i=n}(x - \alpha_i).$$

- **Algebraic (Number) over a Field**

  Let $\mathbb{F} \subset \mathbb{K}$ be fields. $\alpha \in \mathbb{K}$ is said to be algebraic over $\mathbb{F}$ if there exits a polynomial $f(x) \in \mathbb{F}[x]$ of degree $n > 0$ having α as one of its roots.

- **Algebraic Integer**

  A complex number α is said to be an algebraic integer if there exists a monic polynomial (i.e., with leading coefficients unity) $f(x) \in \mathbb{Z}[x]$ of degree $n > 0$ having α as one of its roots.

- **Algebraic Number**

  A complex number α is said to be an algebraic number if there exists a polynomial $f(x) \in \mathbb{Z}[x]$ of degree $n > 0$ having α as one of its roots.

- **Chinese Remainder Theorem** Let $n_1, n_2, \cdots, n_r$ be positive integer such that $gcd(n_i, n_j) = 1$ for $i \neq j$. Then the system of linear congruence

$$x \equiv a_i \pmod{n_i} \quad 1 \leq i \leq r$$

  has a simultaneous solution which is unique modulo the integer $N = n_1n_2\cdots n_r$.

- **Cyclotomic Polynomial**

  Let n be a natural number and $\alpha = e^{\frac{2\pi i}{n}}$ be primitive nth root of unity. Then the polynomial

$$\Phi_n(x) = \prod_{gcd(n,r)=1} (x - \alpha^r)$$

of degree $\phi(n)$ has integer coefficients and is irreducible over the field $\mathbb{Q}$. The polynomial $\Phi_n(x)$ is called nth cyclotomic polynomial.

- **Dirichlet's Theorem on Primes in Arithmetical Progression**

  If a and d are natural numbers such that $gcd(a, d) = 1$, then the set

$$\{a + nd : n = 0, 1, 2, \cdots\cdots\}$$

  contains infinitely many primes numbers.

- **Division Algorithm**

  It is simply the statement that if we divide a nonzero integer by a positive integer then there is some quotient and some remainder. More precisely if a and b be nonzero integers then there exist unique integers q and r such that $a = bq + r$, such that $0 \leq r < |b|$.

- **Epimorphism**

  An epimorphism is an onto homomorphism.

- **Euclidean Algorithm**

  It is name given to the method of computing the Greatest Common Divisor of two integers by the successive use of Division Algorithm. It is based upon the observation that $GCD(a, 0) = a$ if $a \neq 0$ and if $a, b \neq 0$ and $a = bq + r$, $0 \leq r \leq |b|$ then $GCD(a, b) = GCD(b, r)$. We illustrate this by calculating the GCD of 2378 and 1769.

$$\begin{aligned} 2378 &= 1 \times 1769 + 609 \\ 1769 &= 2 \times 609 + 551 \\ 609 &= 1 \times 551 + 58 \\ 551 &= 9 \times 58 + \mathbf{29} \\ 58 &= 2 \times 29 + 0 \end{aligned}$$

  Thus $gcd(2378, 1769) = 29$ (the last nonzero remainder in the above process). We can write $29 = 1769x + 2378y$, where $x = 39$ and $y = -29$. The process of finding such x and y is known as **Extended Euclidean Algorithm**.

- **Euclidean Domain**

  An integral domain D is said to be an *Euclidean Domain* if there exists a mapping

$$\phi :\longrightarrow \mathbb{N} \cup \{0\}$$

satisfying

– Given $a, b \in D$, if a divides b then $\phi(a) \leq \phi(b)$

– Given $a, b \in D$, $a \neq 0$, there exist $q, r \in D$ such that $b = aq+r$, $0 \leq \phi(r) < \phi(a)$

- **Euler's Phi-Function, $\phi(n)$**

For a positive integer n, $\phi(n)$ denotes the number of positive integers less than or equal to n which are coprime to N.

- **Euler's Generalization of the Fermat's Little Theorem**

Let $n \geq 1$ and $gcd(a, n) = 1$. Then $a^{\phi(n)} \equiv 1 \pmod{n}$.

- **Euler's Criterion**

If p is an odd prime and a is any integer not divisible by p then

$$\left(\frac{a}{p}\right) \equiv a^{\frac{p-1}{2}} \pmod{p}.$$

- **Fermat's Little Theorem**

It is the statement that if p is a prime number then $a^p - a$ is divisible by p for all integers a.

- **Fundamental Theorem of Algebra**

Given a nonconstant polynomial $f(x) \in \mathbb{C}[x]$, there exist $\alpha \in \mathbb{C}$ such that $f(\alpha) = 0$. Equivalently if $a_0x^n + a_1x^{n-1} + \cdots + a_{n-1}x + a_n \in \mathbb{C}[x]$, $a_0 \neq 0$, then there exist $\alpha_i \in \mathbb{C}$, $1 \leq i \leq n$ such that

$$a_0x^n + a_1x^{n-1} + \cdots + a_{n-1}x + a_n = a_0(x - \alpha_1)(x - \alpha_2) \cdot \cdots \cdot (x - \alpha_n)$$

- **Fundamental Theorem of Arithmetic**

It is the statement that every positive integer $n > 1$ can be uniquely (apart from the order) expressed as a product of prime numbers.

- **Gaussian Integer**

The elements of the set $\mathbb{Z}[i] = \{a + bi : a, b \in \mathbb{Z}\}$ are called Gaussian integers.

- **Integral Domain**

A commutative ring D with a (multiplicative) identity element is called an integral domain.

- **Legendre Symbol**

Let p be an odd prime and a any integer such that $gcd(a,p) = 1$. Then the Legendre symbol $\left(\frac{a}{p}\right)$ is defined as follows:

$$\left(\frac{a}{p}\right) = \begin{cases} 1 & \text{if a is a quadratic residue mod } p \\ -1 & \text{if a is a quadratic nonresidue mod } p \end{cases}$$

- **Mersenne Primes**

Primes of the form $2^p - 1$, where p is a prime are known as **Mersenne primes**.

- **Minimal Polynomial of an Algebraic Number α**

It is the uniquely determined polynomial $f(x) \in \mathbb{Q}[x]$ satisfying the following conditions:

1. $f(x)$ is monic and $f(\alpha) = 0$.
2. If $g(x) \in \mathbb{Q}[x]$ is such that $g(\alpha) = 0$ then $deg(f(x)) \leq deg(g(x))$.

- **Mobius μ-Function**

Let n be a natural number. The $\mu(n)$ is defined as

$$\begin{cases} 1 & \text{if } n = 1 \\ 0 & \text{if } p^2 \text{ divides } n \text{ for some prime p} \\ (-1)^n & \text{if } n = p_1p_2p_3\cdots p_n \text{ where } p_i, 1 \leq i \leq n \text{ are distinct primes} \end{cases}$$

- **$Ord_a(n)$-Order of a mod n**

If $n > 1$ is a natural number and a is any integer such that $gcd(a,n) = 1$ then order of a mod n is the smallest natural number k such that $a^k \equiv 1 \pmod{n}$.

- **Pigeon Hole Principle**

If $n+1$ items are placed in n (or fewer) boxes then at least one box will contain two or more items. In particular, if x_i, $1 \leq i \leq n+1$ are $n+1$ different real numbers lying in the open interval $(0,n)$ then some two of these numbers will differ by less than 1.

- **Primitive nth Root of Unity**

Given a natural number n, $\alpha \in \mathbb{C}$ is said to be a primitive nth root of unity if $\alpha^n = 1$ but $\alpha^r \neq 1$ for $1 \leq r < n$. For every natural number n, there are $\phi(n)$ primitive roots of unity, where $\phi(n)$ is the Euler's Phi-function.

- **Primitive Root Modulo n**

  Let $n > 1$ be a natural number. An integer x is said to be primitive root modulo n if $x^{\phi(n)} \equiv 1 \pmod{n}$ but $x^r \not\equiv 1 \pmod{n}$ if $1 \leq r < \phi(n)$. A natural number n has a primitive root if an only if the group $\mathbb{Z}_n^*$ is cyclic of order $\phi(n)$. The only natural numbers $n > 1$ having primitive roots are $2, 4, p^k$, and $2p^k$ where p is an odd prime. Generally, the natural number $n = 1$ is also included in this list. If the natural number n has a primitive root then any generator of the cyclic group $\mathbb{Z}_n^*$ is called primitive root mod n.

- **Principal Ideal Domain**

  An integral domain whose every ideal is principal—that is, all of its element consist of multiple of single element.

- **Quadratic Field**

  Let $m \in \mathbb{Z}$ is not a square then the field

$$\mathbb{Q}(\sqrt{m}) = \mathbb{Q}[\sqrt{m}] = \{a + b\sqrt{m} : a, b \in \mathbb{Q}\}$$

  is called a quadratic field.

- **Quadratic Reciprocity**

  Let p and q be two distinct odd primes then

$$\left(\frac{p}{q}\right)\left(\frac{q}{p}\right) = (-1)^{\frac{(p-1)(q-1)}{4}}$$

- **Quadratic Residue and Quadratic Nonresidue mod(p), p an Odd Prime**

  Let p be an odd prime and a any integer such that $gcd(a, p) = 1$. Then a is said to be a quadratic residue mod p if the congruence

$$x^2 \equiv a \pmod{p}$$

  has a solution. If the above congruence has no solution then a is said to be a quadratic nonresidue mod p.

- **Unique Factorization Domain**

  An integral domain in which every nonzero nonunit element can be uniquely expressed as a product of prime elements.

- **A Complete Residue System Modulo n, $n > 0$**

  Let n be a positive integer. A set $a_i \in \mathbb{Z}$, $1 \leq i \leq n$ of *any* n integers is said to be a *Complete Residue System* modulo n if no two of these are congruent modulo n.

- **A Reduced Residue System Modulo n, $n > 0$**

  Let n be a positive integer. A set $a_i \in \mathbb{Z}$, $1 \leq i \leq \phi(n)$ of *any* $\phi(n)$ integers is said to be a *Reduced Residue System* modulo n if each of these a_i is prime to n and no two of these are congruent modulo n.

- **Thue's Theorem**

  Let p be an odd prime and b be any (fixed) integer not divisible by p. Then there exist integers x and y, $1 \leq x, y < \sqrt{p}$, such that $x \equiv \pm by \pmod{p}$ or equivalently $\pm\frac{x}{y} \equiv b \pmod{p}$.

- **Units and Associates in an Integral Domains**

  Let D be an integral domain. An element $u \in D$ is said to be a unit if there exists $v \in D$ such that $uv = 1$, the multiplicative identity of D. Two elements $\alpha, \beta \in D$ are said to be associates if there exists a unit $u \in D$ such that $\alpha = u\beta$. Units in $\mathbb{Z}$ are ± 1, in $\mathbb{Z}[i]$ are $\pm 1, pmi$ and in $\mathbb{Z}[\omega]$ are $\pm 1, \pm\omega, \omega^2$. $1+i$ and $1-i$ are associates in $\mathbb{Z}[i]$ as $1+i = i(1-i)$. $\alpha = \dfrac{-5+\sqrt{-3}}{2}$ and $\beta = 2+\sqrt{-3}$ are associates in $\mathbb{Z}[\omega]$ are associates since $\alpha = \omega\beta$.

Bibliography

Adams, William W., and Larry Joel Goldstein. 1976. *Introduction to Number Theory.* Upper Saddle River: Prentice Hall.

Adler, Andrew, and John E. Coury. 1995. *The Theory of Numbers.* Boston; London: Jones and Bartlett Publishers.

Alaca, Saban, and Kenneth Williams. 2004. *Introductory Algebraic Number Theory.* Cambridge: Cambridge University Press.

Alford, W. R., A. Granville, and C. Pomerance. 1994. "There are Infinitely Many Carmichael Numbers." *Annals of Mathematics* 140:703–22.

Anderson, James A., and James M. Bell. 1997. *Number Theory with Applications.* Upper Saddle River: Prentice Hall.

Andrews, George E. 1989. *Number Theory.* India: Hindustan Publishing Company.

Apostol, Tom M. 1980. *Introduction to Analytic Number Theory.* New Delhi: Narosa Publishing House.

Beiler, Albert. 1966. *Recreations in the Theory of Numbers.* New York: Dover Publications.

Borevich, Z. I., and I. R. Shafarevich. 1966. *Number Theory.* New York: Academic Press.

Bromwich, Thomas John I'Anson. 1991. *An Introduction to the Theory of Infinite Series.* New York: Chelsea Publishing Company.

Bruggeman, T., and T. Gush. 1980. "Nice Cubic Polynomials for Curve Sketching." *Mathematics Magazine* 53: 233–34.

Buddenhagen, J., and C. Ford. 1992. "Nice Cubic Polynomials, Pythagorean Triples, and Law of Cosines." *Mathematics Magazine* 65: 244–49.

Burn, R. P. 1997. *A Pathway into Number Theory.* New York: Cambridge University Press.

Burn, R. P., and Amanda Chetwynd. 1996. *A Cascade of Numbers.* London: Arnold.

Burton, David M. 2011. *Elementary Theory of Numbers.* Wm. C. Brown Company Publishers.

Conway, J. H. October 1973. "Tomorrow is the day after Doomsday." *Engineering Materials and Design* 36:28–31.

Davenport, H. 1993. *The Higher Arithmetic.* Cambridge: Cambridge University Press.

Dedekind, Richard. 1966. *Theory of Algebraic Integers.* Cambridge: Cambridge University Press.

Edwards, H. M. 1977. *Fermat's Last Theorem.* New York: Springer.

Eynden, Charles Vanden. 1987. *Elementary Number Theory.* New York: Random House.

Friedberg, Richard. 1994. *Adventurer's Guide to Number Theory.* New York: Dover Publication.

Gauss, Carl Friedrich. 1986. *Disquisitiones Arithmeticae.* New York: Springer-Verlag.

Gelfond, A. O., and J. B. Roberts. 1961. *The Solution of Equations in Integers.* W. H. Freeman and Company.

Goldman, Jay R. 1998. *The Queen of Mathematics.* Wellesley: A. K. Peters.

Griffin, Harriet. 1954. *Elementary Theory of Numbers.* New York: McGraw Hill Book Company.

Grosswald, Emil. 1984. *Topics from the Theory of Numbers.* Boston: Birkhäuser.

Gupta, Shiv K., and Waclaw Szymanski. 2010. "Cubic Polynomials with Rational Roots and Critical Points." Mathematics Magazine 41: 365–69.

Halmos, Paul Richard. 1981. "Does Mathematics Have Elements?" *Mathematical Intelligencer* 2 (4):147–53.

———. 1991. *Problems for Mathematicians, Young and Old.* Washington: The Mathematical Association of America.

Hardy, G. H., and E. M. Wright. 1960. *An Introduction to the Theory of Numbers.* New York: Oxford University Press.

Harper, Malcolm. 2004. "Z[$\sqrt{14}$] is Euclidean." *Canadian Journal of Mathematics* 56 (1):55–70.

Hsiung, C. Y. 1992. *Elementary Theory of Numbers.* Singapore: World Scientific.

Hurwitz, Adolf, and Nikolaos Kritikos. 1986. *Lectures on Number Theory.* New York: Springer-Verlag.

Ireland, Kenneth, and Michael Rosen. 1990. *A Classical Introduction to Modern Number Theory.* New York: Springer-Verlag.

Jaeschke, Gerhard. October 1993. "On Strong Pseudo Primes to Several Bases." *Mathematics of Computation* 61:915–26.

Khinchin, A. Ya. 1997. *Continued Fractions.* New York: Dover Publications.

Knopp, Konrad. 1961. *Theory and Applications of Infinite Series.* London and Glasgow: Blackie and Son limited.

Koblitz, Neal. 1987. *A Course in Number Theory and Cryptography.* New York: Springer-Verlag.

Koshy, Thomas. 2002. *Elementary Number Theory with Applications.* San Diego: Harcourt/Academic Press.

Kumanduri, Ramanujachary, and Cristina Romero. 1998. *Number Theory with Computer Applications.* Upper Saddle River: Prentice Hall.

Landau, Edmund. 1966. *Elementary Number Theory.* New York: Chelsea Publishing Company.

Leveque, William J. 1990. *Elementary Theory of Numbers.* New York: Dover Publication.

———. 1996. *Fundamentals of Number Theory.* New York: Dover Publication.

Long, Calvin T. 1997. *Elementary Introduction to Number Theory.* Prospect Heights: Waveland Press.

Nagell, Trygve. 1981. *Introduction to Number Theory.* New York: Chelsea Publishing Company.

Niven, Ivan, Herbert S. Zuckerman, and Hugh L. Montgomery. 1991. *The Theory of Numbers.* New York: John Wiley and Sons.

Olds, C. D. 1963. *Continued Fractions.* New York: Random House.

Pollard, Harry, and Harold G. Diamond. 1998. *The Theory of Algebraic Numbers.* New York: Dover Publications.

Rademacher, Hans. 1964. *Lectures on Elementary Number Theory.* New York: Blaisdell Publishing Company.

Ribenboim, Paulo. 1972. *Algebraic Numbers.* New York: Wiley Interscience.

———. 1989. *The Book of Prime Number Records.* New York: Springer-Verlag.

———. 1991. *The Little Book of Big Primes.* New York: Springer-Verlag.

———. 1996. *The New Book of Prime Number Records.* New York: Springer-Verlag.

Roberts, Joe. 1977. *Elementary Number Theory-A Problem Oriented Approach.* Cambridge: The MIT Press.

Rose, H. 1988. *A Course in Number Theory.* Oxford: Clarendon Press.

Rosen, Kenneth H. 1999. *Elementary Number Theory.* New York: Addison Wesley Longman.

Scharlau, Winfried, and Hans Opolka. 1985. *From Fermat to Minkowski.* New York: Springer-Verlag.

Shanks, Daniel. 1993. *Solved and Unsolved Problems in Number Theory.* New York: Chelsea Publishing Company.

Silverman, Joseph H. 1997. *A Friendly Introduction to Elementary Number Theory.* Upper Saddle River: Prentice Hall.

Stanley Ogilvy, C., and John T. Anderson. 1988. *Excursions in Number Theory.* New York: Dover Publications.

Stark, Harold M. 1989. *An Introduction to Number Theory.* Cambridge: The MIT Press.

Stewart, Ian, and David Tall. 2002. *Algebraic Number Theory and Fermat's Last Theory.* Natick: A.K. Peters.

Stillwell, John. 2002. *Mathematics and its History.* New York: Springer-Verlag.

Strayer, James K. 1994. *Elementary Number Theory.* Boston: PWS Publishing Company.

Telang, S. G. 1996. *Number Theory.* New Delhi: Tata McGraw-Hill Publishing Company Limited.

Uspensky, J. V., and M. A. Heaslet. 1939. *Elementary Number Theory.* New York: McGraw-Hill Book Company.

Vardarajan, V. S. 1997. *Algebra in Ancient and Modern Times.* New Delhi: Hindustan Book Agency.

Vinodgradov, I. M. 1954. *Elements of Number Theory.* New York: Dover Publications.

Weil, André. 1984. *Number Theory: An Approach Through History from Hammurapi to Le-gendre.* Boston: Birkhäuser.

Williams, Kenneth S., and Kenneth Hardy. 1996. *The Red Book of Mathematical Problems.* New York: Dover Publications.

Wright, Harry N. 1971. *First Course in the Theory of Numbers.* New York: Dover Publication.

Index